City Publics

As cities have grown into mega-cities or have seen their centres decline with the flight to the suburbs, questions of the public realm and public space within cities warrant even greater attention.

City Publics investigates the ordinary spaces in the city where differences are negotiated. It is concerned with the borders, and boundaries, the constraints and limits on living with, accepting, acknowledging and sometimes celebrating, difference in public. Through ethnographic studies of a number of unusual, surprising and marginal sites, which are not usually the focus of debate, as well as studies of different subjects in public spaces, the book aims to interrogate how difference is negotiated and performed. When and how differences are lived agonistically, and how power is exercised often subtly, not through dominance or manipulation, represents a further focus. Also challenged are the conventional notions of the public and public space.

This valuable and timely book explores the conditions under which violent and negative emotions can erupt to the detriment of others, and elucidates through fine-grained exploration what underlies racist, homophobic, sexist or any other phobic/ist exclusionary practices, so that it becomes possible in some way to expose and confront them. At the same time this book uncovers and reveals some of the many and serendipitous sites of enchantment and pleasure to be found in the city.

With numerous photographs and drawings *City Publics* not only throws new light on encounters with others in public space, but also destabilises dominant, sometimes simplistic, universalised accounts and helps us reimagine urban public space as a site of potentiality, difference and enchanted encounters.

Sophie Watson is Professor of Sociology at the Open University, UK.

Questioning Cities

Edited by Gary Bridge, *University of Bristol, UK* and Sophie Watson, *The Open University, UK*

The 'Questioning Cities' series brings together an unusual mix of urban scholars. Rather than taking a broadly economic approach, planning approach or more socio-cultural approach, it aims to include titles from a multi-disciplinary field of those interested in critical urban analysis. The series thus includes authors who draw on contemporary social, urban and critical theory to explore different aspects of the city. It is not therefore a series made up of books which are largely case studies of different cities and predominantly descriptive. It seeks instead to extend current debates through, in most cases, excellent empirical work and to develop sophisticated understandings of the city from a number of disciplines including geography, sociology, politics, planning, cultural studies, philosophy and literature. The series also aims to be thoroughly international where possible, to be innovative, to surprise, and to challenge received wisdom in urban studies. Overall it will encourage a multi-disciplinary and international dialogue, always bearing in mind that simple description or empirical observation which is not located within a broader theoretical framework would not – for this series at least – be enough.

Global Metropolitan
Globalizing citites in a capitalist world
John Rennie Short

Reason in the City of Difference
Pragmatism, communicative action and contemporary urbanism
Gary Bridge

In the Nature of Cities
Urban political ecology and the politics of urban metabolism
Edited by Nik Heynen, Maria Kaika and Erik Swyngedouw

Ordinary Cities
Between modernity and development
Jenny Robinson

Urban Space and Cityscapes
Perspectives from modern and contemporary culture
Edited by Christoph Lindner

City Publics
The (dis)enchantments of urban encounters
Sophie Watson

Small Cities
Urban experience beyond the metropolis
Edited by David Bell and Mark Jayne

Cities and Race
America's new black ghettos
David Wilson

Cities in Globalization
Practices, policies and theories
Edited by Peter J. Taylor, Ben Derudder, Piet Saey and Frank Witlox

City Publics

The (dis)enchantments of urban encounters

Sophie Watson

LONDON AND NEW YORK

First published 2006
by Routledge
2 Park Square, Milton Park, Abingdon, Oxon OX14 4RN

Simultaneously published in the USA and Canada
by Routledge
270 Madison Ave, New York, NY 10016

Routledge is an imprint of the Taylor & Francis Group, an informa business

Typeset in Times New Roman by
Keystroke, Jacaranda Lodge, Wolverhampton
Printed and bound in Great Britain by
MPG Books Ltd, Bodmin

British Library Cataloguing in Publication Data
A catalogue record for this book is available from the British Library

Library of Congress Cataloguing in Publication Data
Watson, Sophie.
City publics : the (dis)enchantments of urban encounters / Sophie Watson.
p. cm.
Includes bibliographical references and index.
1. Sociology, Urban. 2. Public spaces. 3. Cities and towns–Case studies. I. Title.
HT119.W38 2006
307.76–dc22 2005029954

ISBN10: 0–415–31227–2 ISBN13: 978–0–415–31227–1 (hbk)
ISBN10: 0–415–31228–0 ISBN13: 978–0–415–31228–8 (pbk)

Contents

Illustrations

TABLE

PLATES

FIGURES

Acknowledgements

Friends and colleagues have supported this book in serendipitous ways and I thank them all. Richard Sennett's writing on public space over many years has constantly inspired me. Barbara Caine's generosity, emotional, intellectual and practical throughout the project kept me on the road. Russell Hay's good cheer and interest provided an invaluable support, as did Rosemary Pringle's willingness to step into the breach in Jessie care when needed. John Austin brought older people in the city to my attention. Jessie, my daughter, and Dorothy, my mother, prompted me to consider different concerns and spaces, and I thank them for that. Jeri Johnson introduced me to the delights of Venice in winter where the seeds of this book were sown.

John Austin, Mostafa Gamal, Sandra Koa Wing and Karen Wells provided great research assistance at various points, funded by the National Everyday Cultures Programme at the Open University, and I thank them all. I am very grateful also to Margaret Marchant who helped set up the final manuscript.

Chapter 2 is based on an article of the same name published in *Environment and Planning D: Society and Space* (2005) vol. 23 (4) August, pp. 597–613. Chapter 3 derives from two joint articles I wrote with Karen Wells and I thank her for her contribution. Davina Cooper gave me useful comments on Chapter 2.

I am particularly indebted to Liz Jacka for her most insightful comments on the penultimate draft. Finally I would like to thank Gary Bridge, my fellow traveller in critical urban studies, for making sure this book got finished, and for making collaborative intellectual endeavours such fun.

The author and publishers have made every effort to obtain permission to reproduce copyright material. If any proper acknowledgement has not been made, we would invite copyright holders to inform us of the oversight.

1 Introduction

THE (DIS)ENCHANTMENTS OF URBAN ENCOUNTERS

> I glance around at my fellow citizens as I deposit the books in my sack, and I feel a surge of love for the arbitrariness of our arrangements, that we should be assembled here together in this particular compartment of time, sharing public space, at one with each other in our need for retreat and the printed word. There's Mrs Greenaway, with her impossibly narrow nose bridge, smiling perpetually, an intelligent woman with no place to stow her brand of originality. Mr Atkinson, retired teacher, his tie sunk into the fat of his neck, the *Britannica* opened on the table before him, to a map of some sort. There's a bearded man whose name I don't know but who seems to be scribbling a novel or a memoir into a series of spiral notebooks. There's Hal (Swiftfoot) Scott, who pumps gas and plays hockey, or at least he did before he got caught in a drug bust last year. He's reading *Macleans*, probably the sports section. This is a familiar yet unique scene. The precise patterns will occur only once – us, here, this moment engraved in a layer of memory – a thought that stirs me to wonderment.
>
> (Shields 2002: 45)

For most contemporary city dwellers, or indeed visitors to the city, the experience of walking along a city street, and musing on the diversity of faces they see and languages they hear, on the shops with arrays of different products and smells, restaurants displaying foods and recipes from across the world, is a sensory delight. This is the contemporary phantasmagoric 'multicultural' city, where people of different races, ethnicities, class locations, ages and sexualities live side by side, produced by a complex set of socio-economic, global/local, political and socio-demographic shifts which mean that living with difference, though always a feature of urban life, is probably now quintessentially what city life is about. But running alongside this celebratory urban narrative, constituted by the very same processes, is the city as a space of segregation, division, exclusion, threat and boundaries, where the story of city life as mixing and mingling is replaced by a story of antagonism, fear and exclusion.

These experiences and also myths of the city, in some sense, are not new. Pro- and anti-urban discourses and mythologies have been present since cities first existed. What has changed is the content of the narratives, on the one hand,

and the political and social responses on the other. Thus, in the late nineteenth century, the public spaces of the city were proclaimed unhealthy places populated by the unruly and disorganised working classes, prompting interventions through planning, social reform and other urban strategies. Difference, except as mobilised in discourses of class, was not central to urban narratives, despite the already multicultural composition of many of the world's large cities, as a result of the impact of colonialism and international and inter-regional migration. Even by the time that Simmel was writing, cities were seen not so much as places which concentrated difference but rather as places where sensory overload from excessive stimulation produced a withdrawal and anomie amongst urban dwellers. Meanwhile other writers, such as Tonnies, variously mourned the passing of (imagined) rural cohesive communities (*Gemeinschaft*) for places of loose association (*Gesellschaft*).

What concerns me here is not to find a truth about the living of difference in the public spaces of cities, one overarching narrative or straightforward story. Compelling as such stories may be, evidenced most strikingly in Mike Davis' (1992) depressing description of Los Angeles as the fortress city, articulating the notion that public space has become militarised, or in his more recent depiction of cities as spaces of disaster (Davis 1998) or Mitchell's (2003) portrayal of zones where undesirable behaviour is regulated by what he calls bubble laws, they offer only one account. Similarly, the idea of the end of public space through its privatisation in theme parks (Sorkin 1992), or commercialisation in shopping malls (Crawford 1992), rests on the gaze at some kinds of public space, not others. Rather this book is about public sites out of sight, not the city centres now being designed and planned to reintroduce diversity; it is concerned with the borders and boundaries, the constraints and limits on living with, accepting, acknowledging and sometimes celebrating, difference in public. To argue that public spaces are only spaces of transit where little contact between strangers takes place (Amin and Thrift 2002) is to focus one's gaze on only more visible, and often overplanned, or neglected spaces. There are many other kinds of public space too.

When asked by friends about the book I was writing, I found myself answering that it was about how people rub along, or don't, in the public spaces of the city. This is neither a simple issue nor one to which a universal solution can be found. Moments of tranquillity or harmony can easily erupt into moments of antagonism and violence. Love and hate, empathy and antipathy co-exist in ambiguous and ambivalent tension. Requiring attention, then, are the conditions under which violent and negative emotions can erupt to the detriment of others. Each part of a city is distinct from each other part, and is different at different times of the day and night, as well as across the different months and years, depending on the wider socio-political context. It is also different depending on who you are, both in a material sense and in the realm of the imaginary – every subjectivity in the city is walking through the city streets with a different set of images and imaginations, constituted in personal conscious and unconscious histories. Each city is different from another, though common strands and grounds can be found, and there is a danger in urban studies, all too prevalent, that analyses of American (first), British (next) and other European cities are deployed to describe cities in other parts of the globe, notably Africa, Asia and Latin America, in ways that are utterly inappropriate and even pernicious. It is only when we can elucidate through fine-grained exploration what underlies racist, homophobic, sexist or any other phobic/ist exclusionary practices that we can go some way to exposing and confronting them. Agonistic encounters are an inevitable and productive outcome of differences in the city where these are engaged in with openness and lack of closure, where imbalances of power are acknowledged and addressed, and where outcomes are not pre-determined.

This is also a book which challenges conventional notions of the public and public space – a book which has arisen out of a weariness with the circulation of theories of difference and the public and with well-worn quotations which litter the literature, an irritation with the articulation of assumptions and generalisations about the public space and people within it which are rarely grounded in complex and textured understandings of the people and places concerned. Through ethnographic studies of a number of sites in the city, sites which are not usually the focus of public space debate, as well as studies of different subjects in public spaces, the book aims to interrogate in a fine-grained way how difference is negotiated and lived, when and how differences are lived agonistically, and how power is exercised often subtly, not through dominance or manipulation (Allen 2003), in order to investigate what the limits to difference in different sites at different times might be. Central to my argument is that the specificity and contingency of difference as lived in particular socio-spatial configurations has to be central to urban analysis, even if this specificity illuminates and elucidates wider concerns.

The origin of this book came many years back when I fell in love with public space, that space of delight which encapsulates serendipitous encounters and meanderings: sitting, watching, being, chatting in spaces that may be planned, designed and monumental, but more often may be barely visible to the inattentive eye, on the margins of planned space, or even imagined. It was in Venice that this love affair, which had been bubbling subterraneously for many years, finally erupted. It was not the grand Piazza San Marco which charmed me, as it has charmed many urban designers and planners before, despite its awesome beauty. Campo Santa Margherita stole my heart.

This is a public space which is irregular, haphazard and ordinary. Its ten entrances/exits invite random paths to be taken, its benches, scattered across the square, lure the old and young to pause for a while, its lack of cars entices kids to play and chase the pigeons, its market stalls bring locals to shop, its calm and bustle, light and shade, mark it as a place to gaze, chat and rub along with others with ease.

The notion of random, specific, contingent, symbolic, imagined and lived, visible and invisible, spatio-temporally differentiated public space informs the selection of sites and subjects deployed and examined here. These are not Richard Rogers' grand piazzas or endlessly rehearsed shopping malls. A vignette from the month before this book went to press illustrates the point. On a hot day in the early

Plates 1.1 and *1.2* Saturday morning, Campo Santa Margherita, Venice. Photographs: Jeri Johnson.

summer, having failed to find the promised fun fair, my daughter, her friend, my friend and I set off to another borough through the bank holiday traffic to a planned children's play area in a local park. The visit was disastrous. The place was crowded, my child was wearing the wrong clothes to play in, the two competed for the equipment and showed off to each other, parents were grumpy, the place was littered and ugly. We beat a hasty retreat with the children in tears in the back of the car. Attempting to retrieve the situation, we bundled them off to the local city farm, a space cut out of the railway sidings and abandoned land, captured from the railway authorities by a local community group, where in a higgledy-piggledy 3 acres, horses, cows, goats, sheep, a pig and chickens share the space with tumbledown buildings, an education centre, stables, a couple of fields, allotments for old age pensioners and, crucially on this particular day, a pond. There, by the side of this small pond, were children of all ages, ethnicities and class backgrounds, lying on the ground fishing for tadpoles in plastic cups, while parents sat and lay on the banks chatting. It was two hours before any of the kids could be extracted from this buzzing, intermingling, cheerful site to return home. In this scruffy, unplanned and marginal public space, on this particular afternoon, urban encounters across age, race, sex and class enchanted and surprised those who happened upon it. This is London, a global city, but in every city in every corner of the globe there are sites of magical urban encounters, hidden in the interstices of the planned and monumental, divided and segregated, or privatised and thematised, spaces that more usually capture public attention.

A number of themes and arguments run through this book, foregrounded differently across the sites and subjects here. The first is that there are different

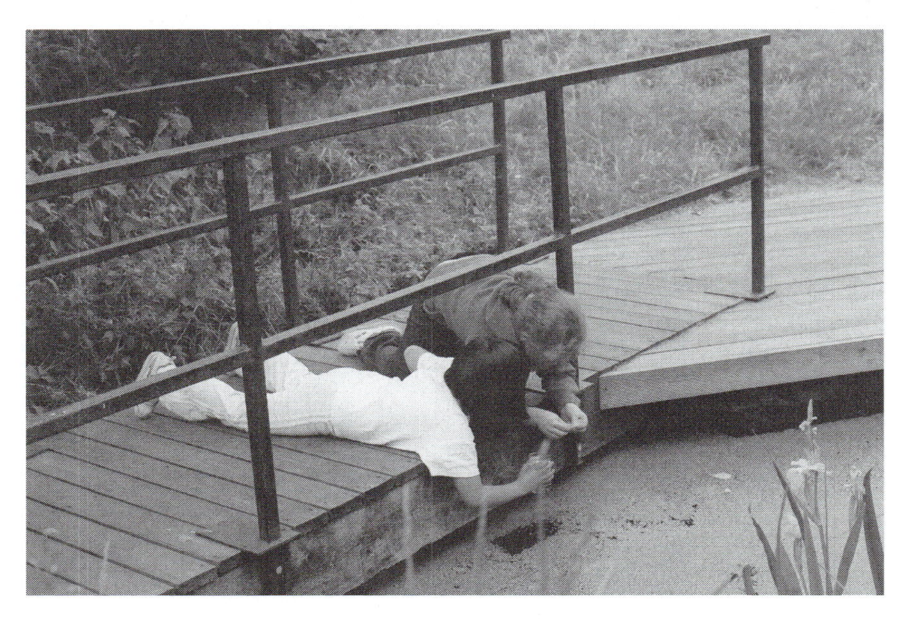

Plate 1.3 Children search for tadpoles at Kentish Town City Farm, London.

conceptions of the public for different subjects: the 'public' and public space are deployed and understood in multifaceted and particular ways, constructing subjects heterogeneously. The second is that *contra* Habermas and others, the public is not just about 'talk', it concerns bodies and their micro-movements. To put this another way, bodies and public space are mutually constitutive. The third argument is that exploring marginal, unportentous, hidden and symbolic spaces, and the different imaginaries of often forgotten subjects, gives us a way into thinking of public space differently. At the same time, connecting to a third argument, these very same sites can quickly shift from liminal space to centre stage as they rub up against institutional and regulatory arenas at particular historical moments. So even marginality is a temporary and shifting state: the invisible becomes visible and vice versa in unpredictable ways. Thus, the role of the state and regulation in constructing public space is another theme in the book. The fourth theme is the mutually constructed and complex relation between public and private spaces, and the culturally embedded nature of embodied and social practices associated with these. Fifth, this is a book about engaged, agonistic and antagonistic encounters in the city – an exploration of how and when difference matters. Central to all these themes is the question of how different subjectivities, bodies and knowledges are constructed in, and themselves construct, public space.

For me, this is an important political project. The city, and the public spaces which constitute it, in the twenty-first century is the site of multiple connections and inter-connections of people who differ from one another in their cultural practices, in their imaginaries, in their embodiment, in their desires, in their capacities, in their social, economic and cultural capital, in their religious beliefs, and in countless other ways that cannot be enumerated here. If these differences cannot be negotiated with civility, urbanity and understanding, if we cling to the rightness of our own beliefs and practices, and do not tolerate those of another in the public spaces of the city, at best Mike Davis will have been proved right, at worst there will be no such thing as city life, as we know it, to write about or celebrate. This is not to argue for a world where differences are ironed out, equalised or placated. As Mouffe (2000) persistently contends, differences are inherently agonistic and, following Foucault, implicated in the exercise of power in complex and productive ways. Rather, we have to confront the realities of differences in the city head on, experience the pleasures as well as the pains they inevitably produce, and think about ways of not entrenching difference in the politics of fundamentalism (religious and secular) which is threatening the possibility of everyday democratic multicultural space. So, finally, to borrow Bennett's (2001) notion, this also is a story of enchantment in the public spaces of the city, a story of places and sites where people do rub along, not just in the exoticised, celebrated and commodified spaces representing a visible multicultural settlement, so loved by city planners and investors, but in the ordinary spaces of everyday life.

In the rest of this chapter I introduce some of the theoretical propositions that have informed my thinking. I then consider different conceptions of public space and the public realm. The first part of this discussion explores conceptions which foreground the potentialities for the co-mixing and mingling of strangers in public space or which see public space as the realm of debate, citizenship and democracy. The second part considers different explanations for the limits to living with difference in the city, which help elucidate some of the stories which follow.

CITY PUBLICS: THE (DIS)ENCHANTMENTS OF URBAN ENCOUNTERS

This book, then, is about encounters in public urban space. It is also about difference and the multicultures that inhabit the everyday spaces of the city, and the ebbs and flows of openings and closures, the possibilities and constraints, the inclusions and exclusions, the joys and pains, that these produce. Public space is always, in some sense, in a state of emergence, never complete and always contested, constituted in agonistic relations, in that it is implicated in the production of identities as relational and produced through difference. As Connolly (1998: 93) puts it:

> Public space is the space in which such collaborations and contestations occur. Today, a profound source of social fragmentation flows from the demand by a series of intense contenders to occupy the authoritative center. That center must itself be pluralised.

This represents a break with the idea of public space as a space where differences as fixed identities can be asserted. Rather, following Deutsche (1999: 176), it 'interrogates exclusionary operations that simultaneously affirm and prevent the closure of identity': opening it up instead as a space of heterogeneity where differences are acknowledged as constituted in power relations.

Weber places disenchantment at the centre of modernity in his description of the modern, rational, routinised, bureaucratised and secularised society (Jenkins 2000) where life is increasingly impersonal and mystery and magic have no place. These ideas have permeated political thought as well as commonsense understandings of everyday life. Notwithstanding the usefulness of Weber's insights, like Bennett (2001) in a different context, I want to reintroduce an idea of enchantment into our sense of living together, with all the fractured, fluid, shifting and different subjectivities implied, in the spaces of the city. In part, then, this book is an attempt to reclaim public space from the darker narratives which have constructed it over recent decades. Stories of the city and its public spaces as dangerous, dead or dull, or as sites of exclusion, marginalisation and violence, I want to suggest, contribute to, and produce, the very conditions that they describe. Some accounts of this kind do appear here. But new stories of public space as life enhancing, exciting, safe and inclusive, or stories of sites that are on the margins and barely visible, can take us far in creating those spaces in just that way. When word goes out that a market or a new walkway by a river, which days before was empty, is fun to visit, crowds are drawn there, creating the very sense of co-mingling with others – the delights of encounter – in the city that those who go there seek. This in turn attracts street performers, food or second-hand book stalls, jewellery stalls, bric-à-brac, sights to surprise and charm, sounds and smells to arouse, and the random occurrences of everyday public space.

Representational space (Lefebvre 1991), or space as psychologically lived in, is implicated in the very production of space itself. How we imagine a place, space, city in large part creates the conditions of possibility for how we act, which itself creates the contours of that very space. The stories we tell ourselves as we walk down the street, the swirling of affect, the cacophony of noise, take us along one route or another, down this alleyway or highway, to this park, to this market and this street. Bombarded daily by images and stories of public space as dirty, polluted, dangerous, empty, homogenous, pointless – the list is endless – we retreat into the private realm of family and friends, or the privatised realm of cultural consumption that we can afford. To protect ourselves from meeting and confronting those who threaten or disturb us, who disrupt our sense of self, who put our fragile subjectivities under threat, we draw into ourselves in indifference at best or hostility at worst. But this presumes the possibility of such a withdrawal from others. Jean Luc Nancy would have us see it otherwise:

> we happen – if happen is to take place, as other in time, as otherness. . . . We are not a 'being'. . . . This happening as the 'essential' otherness is given to us as we, which is nothing but the otherness of existence. . . . The 'we' is not, but we happen, and the 'we' happens, and each individual happening happens only through this community of happening, which is our community. . . . Community is finite community, that is, the community of otherness, of happening . . . history is community, that is, the happening of a certain space of time – as a certain spacing of time, which is the spacing of a 'we'.
>
> (1993: 156)

For Nancy, then, a person exists, or has a singular history, only insofar as he or she is exposed to and within a community – the '"we" happening as the togetherness of otherness' (ibid: 158). This is a thoroughly situated, in space, time and place of otherness, sense of self and subjectivity, formed in the public spaces we inhabit with strangers – rubbing along, as I have chosen to call it, to underline its very ordinariness.

The challenge of our time is to conceive of society and political life as having difference at their heart, rejecting the notion of multiculturalism as deployed by a dominant group to spatially manage those that are other (Hage 1998). Multiculturalism which rests on a notion of cultures as definable and homogenous is inevitably a view from the outside used as a means to understand and control others from a space of power (Benhabib 2002: 8). Cultures are not fixed and given, but fluid, shifting and contested. So too are identities. As Benhabib puts it:

> human identities can be formed only through webs of interlocution . . . the individual can be seen to have a 'right' – that is, a morally justifiable claim of some sort – to the recognition by others of structures of interlocution within

> which he or she articulates an identity, only if it is also accepted that each individual is equally worthy of equal treatment and respect.
>
> (ibid: 56)

There can be no normative assumptions as to which collective life forms, in particular which ways and strategies of inhabiting public space, should be privileged over others. Here I am also compelled by Wendy Brown's (2005) rejection of the notion of tolerance, which, she argues, works to mask the universalism at the heart of liberalism and disguises its normative powers. Central to liberalism is the autonomy of the individual subject, on the one hand, and, on the other, the autonomy of law and politics from culture, thus securing the individual from culture, which is depoliticised and relegated to some other place into or from (parts of) which the individual can voluntarily enter or leave. Thus tolerance configures the right of the liberal subject to tolerate others or not, and is 'generally conferred by those who do not require it upon those who do'. Only those who deviate from, rather than conform to, established norms are thus eligible for tolerance.

Strategies and discursive practices deployed by dominant groups to marginalise those less powerful in the representation, definition and use of public space rely on a self-perception which is produced and reproduced with such subtlety as to deny their normative effects. William Connolly (1995) makes the point that one of the difficulties with the growing pressure towards pluralisation is that the opening up of a cultural space for the redefinition of difference presupposes a redefinition of the terms of self-recognition on the part of those whose community is being pluralised. It is for this reason, he argues, that 'the pressures to pluralization and fundamentalization so readily track one another' (ibid: xvii). Castells (1997) makes a similar kind of point when he argues that the rise of Christian fundamentalism has taken place in the context of progressive movements for women's and gay liberation in the USA. Central to Connolly's argument is the idea of critical responsiveness, which means revising the terms of one's own self-definition and modifying the shape of one's own identity; also a respect for, and recognition of, new identities even as they are in the process of becoming, that is before they take shape around a stable definition that can be recognised. Identities in this framework are always relational and collective – an argument which runs through all of his work (including Connolly 2002). Connolly rails against the reduction of the notion of difference, which is relational, to diversity, which implies independent identities; in other words, 'there is no identity without difference' (1995: xx). Again, like Castells, Connolly emphasises that the more forceful the push towards integrated communities the 'more implacable the drive to convert difference into otherness' (ibid: xxi). What this implies in terms of public space, I think, is an openness and porosity, a blurring of the boundaries, a permeability where differences can collide and rub up against each other, even while the other is recognised as different. Though there are arguments for separate spaces for different groups, particularly where imbalances of power inhibit the expression of specific cultural/gendered/sexed/raced practices, where these are solidified into bounded identity communities, relationality and encounter are lost.

The move whereby 'identity converts difference into otherness to secure its own self-certainty' (Connolly 2002: xiv) is thus a dangerous one, if self-certainty means recourse to the implacable and immovable subject position.

Connolly's political and philosophical project, then, is to construct a new ethos of political engagement and connection between people which is founded in part on the notion of 'agonistic respect' which he describes as:

> a civic virtue that allows people to honor different final sources, to cultivate reciprocal respect across difference, and to negotiate larger assemblages to get general policies. . . . Agonism is the dimension through which each party maintains a pathos of distance from others with whom it is engaged. Respect is the dimension through which self-limits are acknowledged and connections are established across lines of difference.
>
> (ibid: xxvi)

Coming together across their multiple differences, he proposes a generous ethos of engagement where different identities and faiths are appreciated and articulated in debates and decisions about the fundamental issues of public life (Connolly 1998: 94).

My concern, like Connolly's and also Deutsche's, is how to conceive of democratic public space which is not predicated on the exclusion of those who are different from ourselves. Drawing on Emmanuel Levinas, Deutsche (1999: 176–185) cites the notion of 'non-indifference to the other' (Levinas 1994: 124, quoted in Deutsche 1999: 176), or to put this another way, taking 'the other' seriously. For her, indeterminacy – the abandonment of references to a transcendent ground of power – 'exposes us to others, and with exposure, democracy is invented' (Deutsche 1999: 184). For if the grounds of our commonality are uncertain, this opens up the space for continuing debate and negotiation around social questions and rights. To quote: 'The removal of its [democratic society's] ground pluralises society – not by fragmenting it into self-contained, conflicting groups, but by making it incompletely knowable and therefore "not mine"' (ibid: 185). There is an idea here of lack of fixity opening up the grounds for debate and dismantling the prior claims of strong and homogenous groups to define the terms and to claim power. Public space in these terms becomes the space where these debates and negotiations can be enacted, allowing for the possibility that different claims will be made at different times by different emergent groups. Following Levinas' idea that the reasonable human being can be defined by his or her non-indifference to the other, 'democratic rights can then be understood not simply as the freedom *of* the self but a freedom *from* the self, "from its egotism"' (Deutsche 1999: 185). This gives a breath of fresh air to the centrality of self and the individual, so fundamental to contemporary political, economic and social discourse.

In my own reading of Levinas, I am attracted again, as I am with Connolly, by his appeal to an ethics of conduct between strangers (not kin) who encounter one another: 'All thought is subordinated to the ethical relation, to the infinitely other in the other person, and to the infinitely other for which I am nostalgic. Thinking the other person is a part of the irreducible concern for the other' (Levinas 1999: 98). How to construct cities as spaces of difference where tolerance does not mean the power of one group to recognise, embrace and include the other (or not), with all the implied power imbalances and fixities of identity and position, is a key concern in this work. What is implied is thinking about new forms of ethical conduct which have the notion of difference and otherness at their core.

Having laid out these ethical and political notions for thinking about encounters with those who are different from ourselves in the urban public space, I want now to move on to look at some of the dominant conceptions of public space which have informed my thinking to date. In Part 1 I discuss those theories which provide a relatively optimistic reading of the possibilities of encounter in public space. Part 2 explores different understandings of the limits to public space as a site of encounter.

CONSTRUCTING PUBLIC SPACE

Part 1: Public potentialities

In *The Human Condition* Hannah Arendt (1958) argues for an idealised public realm which draws on the idealised agora of ancient Athens where people came together to talk freely as non-economic selves, away from the life of labour which has the capacity to oppress them. For Arendt (ibid: 11–12) the significance of the public realm lies in the fact that it is a space away from family life where everybody sees and hears from a different position and things can be seen and heard from a variety of perspectives and aspects. To quote Arendt (ibid: 12): 'The end of the common world has come when it is seen only under one aspect and is permitted to present itself in only one perspective'. This then was a space of heterogeneity. Arendt recognised that people are not born equal and as such she was highly sceptical 'of all those tendencies in modern life that foster a false sense of social equality and homogeneity' (Bernstein 1996: 86). Thus she argued that to deny or iron out differences among human beings was also to deny their distinctive individuality; instead differences were to be recognised and celebrated, thus enabling a viable public realm and political community to exist. So in her account the public realm becomes a space of appearances where individuals perform great deeds and speak memorable words (Villa 2001: 244). Two models are constructed in her work, one where the public sphere is theatrical, performative and agonistic, the other where it is more associational, leading to a democratic politics engaged in by ordinary citizens (ibid). For Benhabib the theatrical model closely corresponds to Greek experience which presumes a high degree of moral and political homogeneity, while the associational model of public space is only viable under conditions of modernity. The permanence of public space was also important for Arendt (1958: 55), the idea that it must be created to transcend the life of mortal man, since without this, for her, 'no common world and no public realm is possible'. This is a very different conception of public space from more recent ideas of it as contingent, fleeting and fluid.

Habermas, like Arendt, proposes an idealised public realm as a space of communication, though as Sennett has pointed out (2000: 383), this is in some respects a richer version since he reintroduces aspects of civil society, notably labour and economic life. Central to Habermas is the notion of impersonal communication where interests are talked about and debated rather than defended, and where consensus is reached through rational debate. This was why he chose to emphasise the coffee house as a place of talk and newspaper reading and thus as a key site of the public realm, essentially demarcated as the bourgeois public sphere. In his account there is a clear division between private and public realms; as people enter the public realm they leave behind their private interests to engage in rational political discussion around common concerns. With the growth of the welfare state on the one hand and the mass media on the other, this public realm is increasingly threatened as private–public boundaries break down and the public realm is more and more inhabited by people who do not come from the bourgeoisie. Feminists, amongst others, have critiqued Habermas' account as predicated on strong divisions between public and private spheres and on an open debate based on critical reason which is, in fact, assigned exclusively to the masculine subject.

Drawing on a different set of traditions, Iris Young (1990), whose work is often cited mantra style by urbanists, also proposes an optimistic reading of how cities can be the site of the co-mingling and encounter of strangers, who are able to express and perform their differences in proximity without interaction or discourse being necessary. Young's starting point is a critique on two grounds of dominant notions of community. The first is that they deny difference within and between subjects relying on the Cartesian understanding of subjectivity, basic to the modern metaphysics of presence, where self-knowledge, knowledge of another and mutual understanding are crucial. Second, they are predicated on face-to-face relations in decentralised, often economically self-sufficient small communities. These 'authentic' social relations are opposed radically to those produced in a larger alienated society. As Young puts it, these social relations 'are not mediated by space and time distancing'. Rather than seeking to dismantle the city, Young seeks to embrace and celebrate modern urban life in all its complexity, its multiplicity of activities and people, and its capacity to embody difference. The fact that in the city strangers live side by side in public spaces instantiates social relations as difference. This is 'the unoppressive city . . . defined as openness to unassimilated otherness', and a 'politics of difference [which] lays down institutional and ideological means for recognising and affirming differently identifying groups' both through the granting of political representation to different group interests and through celebrating their different cultures (Young 2002). These aims obviously raise important and difficult questions and a challenge to move beyond the conventional pluralism of liberal interest groups, which are not my concern here. I have drawn particular attention to Young's ideas partly because they have been so influential, partly because I have great sympathy with them, and most centrally because this book seeks to interrogate some of the limits and barriers to the realisation *tout court* of Young's ideals of city life.

In Jane Jacob's *Death and Life of American Cities*, we find another influential vision of the city's potential. Here the sidewalks play an active part in the 'drama of civilisation versus barbarism' in cities: when filled with strangers moving along together, the streets become safe and enticing places to co-mingle. For her the essence of good city planning is to balance people's desire for privacy with their desire for differing degrees of contact with one another. Such a street has a clear demarcation between private and public and 'the more strangers the merrier'. Arendt, Habermas, Jacobs and Young, it could be argued, all in different ways propose an idealised notion of public space which has the capacity to enable the co-presence of people different from each other either in debate or simply in a mutually productive proximity where each other's difference is recognised and acknowledged. De Certeau (1984) in a different way posits the public spaces of the city as sites of potentiality. For him, simple everyday spatial practices like walking in the city can work against the quotidian discipline of the rationalist model. In the same way that speech acts can make new meaning within the disciplinary confines of language, so also walking in the city has a rhetorical function in that it can weave together different elements of the city's possibilities – meandering or taking short cuts – and in such unplanned movements make different meanings. So walking, like speech acts, has an enunciative function; its mobility and paths through the city cannot be constrained by disciplinary power (ibid: 337). Thus for de Certeau everyday practices have the capacity for resistance, creating new public spaces drawing on the heterogeneity of the city. Many other writers, urban designers and practitioners have also variously shared these ideals – notably Richard Sennett, as we see shortly – while being more attentive to the constraints and forces which limit its actualisation in real city spaces. In this next section I turn to different understandings of how and why the city, in particular the capitalist city, falls short of these ideals.

Part 2: The limits to difference

In *Metropolis and Mental Life*, Georg Simmel (1903) provides a wonderful account of how the modern city, with its continuous bombardment of the senses through external and internal stimuli, has produced a new metropolitan individuality. Differences lie at the heart of his argument – our existence is dependent on them. Rapidly changing images and impressions, produced in the modern metropolis each time we walk down the street, with its multiplicities of social, economic and occupational life, has produced an urban consciousness which is in striking contrast to that of the slower, flowing and more habitual mentality of the rural or small town dweller. In order to protect him/herself against the plethora of sensations and stimuli he/she encounters, the metropolitan individual develops a protective layer against the disruptions and discontinuities that threaten him/her. Thus, Simmel argues, 'the reaction of the metropolitan person to those events is moved to a sphere of mental activity which is least sensitive and which is furthest removed from the depths of personality' (ibid.: 325). Mediated thus by intellect and by money as the form of exchange between people, impersonal and indifferent relations between them become the norm. So too the excess of new stimulations and an inability to react to these produces in the city dweller a blasé attitude which itself generates an indifference to distinctions between things, which become flat, colourless and homogeneous as a result. In the midst of this 'the mental attitude of the people of the metropolis to one another may be designated formally as one of reserve' (2002: 15), whose inner side 'is a slight aversion, a mutual strangeness and repulsion, which, in a close contact which has arisen any way whatever, can break out into hatred and conflict'. This analysis of the psychic life of the city dweller calls to mind contemporary patterns of behaviour which are so much a part of city life – road rage, extreme reactions to mild events of no consequence like an incidental touch from a stranger passing by in the street, and constant verbal abuse. These hostile practices, I suggest, are principally produced by excess stress and overload in urban space.

Walter Benjamin (1999) pays similar attention to the psychic effects of the modern metropolis but comes to different conclusions. For Benjamin, the city is a place of pleasurable sensation, a phantasmagoria of bright lights, new commodities and fleeting and intoxicating impressions. In *The Arcades Project* we are taken on a dizzy ride past shops selling combs of many colours, umbrellas, stockings as we rush on along mysterious alleyways and up spiral staircases rising into darkness (ibid: 396). This phanstasmagoric world reaches its height in the world fairs which Benjamin sees as the 'folk festivals of capitalism' (Buck-Morss 1989: 220). In the *Passagen-Werk* Benjamin is concerned with the effect of these fairs on the working classes, whose message 'as fairylands was the promise of social progress for the masses without revolution. Indeed the fairs denied the very existence of class antagonisms' (Buck-Morss 1989: 220). Then later, following Haussmann's large urban renewal projects in Paris, which carved the grand boulevards through the city, there is a new social utopia where the revolutionary potential of the working classes is once more undermined and the people no longer feel at home in the city. So Benjamin's capitalist city is not so much one where people withdraw into themselves for protection from the effects of encountering strangers, but one where delight is to be found in the commodified city where class is the crucial social division, both co-opted and managed, but not eliminated.

Over the last 30 years or so, Richard Sennett (1970, 1974, 1990, 1994) has been the urbanist to defend most emphatically the notion of public space as crucial for broader notions of democracy, urban civility and the public sphere. A great admirer of Arendt, he has drawn on her writings to argue that the public realm can be characterised by an idea of the richer types of relationships that are possible amongst strangers. Upturning the trope of the family and the private realm as the site of intimate and fulfilling relationships away from the anonymity of the city, a dominant (conservative) political discourse, Sennett (1990, 2000) argues that if an idea of the complexity and richness of impartial relations in public is lost, the realm of politics becomes that of assumed intimacy and the cult of personality. Similarly if the open and unpredictable spaces of the city are lost, the potential for sociality in the city is lost also. In contrast to Simmel, then, under-stimulation, not over-stimulation, has mobilised a retreat from the city. This argument runs through all of Sennett's books, but most famously is elaborated in *The Fall of Public Man* (1974), where he explores how the forces of industrial capitalism, with the pressures of privatisation and secularism, began to erode public life in the nineteenth century. By the late twentieth century, with the flight of the middle classes to the suburbs, in the USA particularly, and urban interventions in planning and architecture, the public spaces of the city, Sennett suggests, had become dead, offering fewer and fewer opportunities for unpredictable encounters amongst strangers in public. As a result people have retreated further and further into intimate and private relations with others. In his later work, Sennett proposes various ways out of this dilemma. In the arena of urban design he argues for the importance of complexity, surprise and disorder in creating spaces conducive to spontaneous co-mingling of strangers. These more multifunctional spaces of the city offer opportunities for role playing and performativity, a theatre of public life – a *teatro mundi* – which by 'design is more attuned to the question of difference and make[s] the question of difference concrete' (Sennett 2000: 385).

There is a danger, though, in idealised versions of a lost public realm, which runs through these accounts. Deutsche (1996) draws our attention to it in her exploration of the place of public art in cities. For as she and many feminists in different contexts point out, space is inherently conflictual and implicated in struggles over inclusions and exclusions. Women and slaves had no part to play in the Athenian agora, which leaves us with the question: whose loss are we mourning here? This long history of urban thought which emphasises the loss of a once vibrant city life has taken a new turn in much American urban literature in particular, towards a doom and gloom discourse of urban life and public space. In part it is a desire to destabilise and interrogate these rather pessimistic and universalising accounts, and to look at some specific sites and modes of interaction in ordinary everyday public spaces in their finer-grained texture which has prompted this book.

Yet sensationalist accounts sell books, so it should be no surprise that Mike Davis' (1992) *City of Quartz* has sold more widely to the general public than any other urban text since about 1990. Here he paints a picture of Los Angeles as a fortress and militarised city, where street benches are constructed in a way that prevents people sleeping on them, where sprinklers in the park deter the homeless at night, where shopping centres and libraries are patrolled and surveyed to keep out those deemed undesirable, and where floodlights from the LAPD helicopters scan the streets at night. This narrative is further embellished in his more recent apocalyptic account of *Dead Cities* (2002), the combat zone of urban America, beset by white flight, de-industrialisation, housing and job segregation, and discrimination. Here we find tales of urban neglect, greed and political scandal, and urban infrastructures collapsing from environmental disasters quite as potent as any terrorist threat.

For Mitchell (2003), this increasing fortress mindset, particularly post 9/11, has mobilised new strictures on behaviour. Bubble laws, as he describes them, regulate the movement of protesters in public space (ibid: 44), and homeless people are seen as threatening and out of place in public space and are therefore moved on (ibid: 183). For Neil Smith (1996) too, the twentieth-century discourse of urban decline in the USA, which Beauregard (1993: xi) argued functioned ideologically to produce certain kinds of response, has now become a material reality in the de-gentrified, revanchist city, where race/class/gender terror is:

> felt by middle and ruling class whites who are suddenly stuck in place by a ravaged property market, the threat and reality of unemployment, the decimation of even minimal services, and the emergence of minority and immigrant groups as well as women as powerful actors.

Here we have the empire striking back as whites try to leave the city to those they have exploited. For Sorkin the new public spaces are a theme park from which he looks nostalgically back to the 'familiar spaces of traditional cities, the streets and squares, courtyards and parks' (1992: xv). In the last decade of the twentieth century, these were the dominant narratives of the US city which repudiated everyday space as a space of possibility as articulated in the more idealistic visions of Arendt, Young, Jacobs and others. There has been a striking paucity of studies of the mundane and commonplace spaces of the city where people simply muddle through or rub along, living and performing their differences, and even delighting in them.

PSYCHOANALYTIC ACCOUNTS

Accounts of the decline of public space are one part of the story. The other is the people who inhabit them. The city has always been the site of complex and intense interconnections and juxtapositions (Allen *et al.* 1999), the site of living with difference. This indeed defines city life and is what makes it distinctive and, historically at least, different from rural or suburban living – though these distinctions are increasingly difficult to uphold. In the last three or so decades of the twentieth century the combination of social demographic and globalisation processes meant that cities became more multicultural than ever before. On the one hand changing social mores, as a result partly of the feminist and gay liberation movements from the late 1970s, partly of growing numbers of people in higher education and partly of wider access to different ideas through music and television, have produced a huge diversity of households, including larger numbers than ever before of single and gay households. On the other hand, in many cities migrants from other countries now nearly equal or out-number people born in the locality. Ethnic and racial differences are now integral to the fabric of life in cities across the globe. Increasingly, city dwellers and visitors to the city are encountering those who are clearly different from themselves, visibly as well as socially and economically.

Notwithstanding the political and institutional structures which shore up the dominance of certain cultural groups, how can we make sense of resistance to difference in others? Psychoanalytic accounts are useful here. The French psychoanalyst Julia Kristeva gives an insightful archaeology – as she describes it – of attitudes to the foreigner. For her, the place of the foreigner in our psychic life is integral to our sense of self: 'Strangely the foreigner lives within us: he is the hidden face of our identity, the space that wrecks our abode, the time in which understanding and affinity founder . . . by recognising him within ourselves, we are spared detesting him in himself' (1991: 1). Thus by splitting off the part of our selves that we fear and detest, the foreigner within – a process central to Freudian theory – we avoid the things we cannot face. According to Kristeva the foreigner can only be defined in a negative fashion (ibid: 95). This produces in the foreigner a demeanour of indifference and aloofness as a shield and protection against the attacks and rejections that he experiences. For Kristeva there is a paradox: 'without a social group structured about a power base and provided with legislation, that externality represented by the foreigner and most often experienced as unfavourable or at least problematical would simply not exist' (ibid: 96). Drawing on the work of Daniele Lochak, Kristeva sees the foreigner as signifying:

> the difficulty of living as an *other* and with others; politically he underscores the limits of nation-states and of the national political conscience that characterises them and that we all have deeply interiorised to the point of considering it normal that there are foreigners, that is, people who do not have the same rights as we do.
>
> (Kristeva 1991: 103)

What is needed in the final analysis, in her view, is for us to extend to the notion of the foreigner the right of respecting our own foreignness and, in short, the '"privacy" that insures freedom in democracies'.

David Sibley is probably the geographer who has developed psychoanalytic ideas most fully in order to understand contemporary practices of exclusion – notably the exclusion of gypsies. Drawing on object relations theory developed by Melanie Klein, he describes how:

> the self and the world are split into good and bad objects, and the bad self, the self associated with fear and anxiety over the loss of control, is projected onto bad objects. Fear precedes the construction of the bad object, the negative stereotype, but the stereotype – simplified, distorted and at a distance – perpetuates that fear.
>
> (Sibley 1995: 15)

Stereotypical views of people in a conflict situation arise when a community that represents itself as the norm feels threatened by those whom they perceive as different and threatening. The fear and anxiety are translated into a stereotype (ibid: 28–29). Though engaging with the threatening other might dissolve some of this fear, as Sibley points out, a limited and superficial engagement with those who are different might be even more problematic than none at all if it produces only limited knowledge and understanding. These are the processes which construct group boundaries to keep those who threaten them at a distance (ibid: 46). In a similar vein Stallybrass and White (1986: 191) suggest that the desire for differentiation from others depends upon disgust.

THE BOOK'S OUTLINE

It should be clear by now that I do not aim here to construct a grand narrative of public space. Nevertheless, I want to argue for an understanding of public space and the encounters enacted there which is predicated on an ethical commitment to the public acknowledgement of others who are different from ourselves, to the agonistic relations these may produce, and to the ongoing debates, arguments, resolutions, shifts and turns that will inevitably result. Relations of difference are always implicated in power. Their potential not to oppress, exclude or subjugate others can only be realised when the homogeneity and power of dominant groups, and their defining and othering discourses, are themselves disrupted, destabilised and challenged. This will also mean addressing the historical, socio-cultural and political specificity of each and every public space and the cultural practices taking place there.

This book thus does not aim to tell one story of public space, or to construct one overarching account. Rather, it aims precisely to do otherwise. Taking unusual sites and spaces, not the coffee houses, piazzas, malls, city centres or theme parks around which so much public space discourse has been debated, I make visible sites and spaces that are symbolic, ordinary and sometimes invisible. Similarly, the subjects of this book are not selected to represent difference and otherness in any predictable way. On the one hand, white Englishness is interrogated and exposed, particularly through the sites selected. So too, gender, sexuality, race/ethnicity and class, and the ways these are represented and performed in public space, are read through these sites. But also in the book, groups of people who are themselves often barely visible in public space – older people and children, find a voice.

I share with Amin (2002: 2) an emphasis on the micro-publics of social contact and encounter which provides us with an understanding of ethnicity, and other identities too, as a mobile and incomplete process. The ethnographies in this book have thus been selected to illustrate some of the ways that differences are lived in the public spaces of the city, always bearing in mind that there are many more and that no story of these can ever be complete. So across the chapters there are encounters with difference where these are productive and positive, while at the same time I attempt to expose some of the limits to difference in the public spaces of the city. Sometimes these are simply a result of othering discourses and exclusions, deriving from a sense of power and authority amongst a dominant social group – usually the longer-term white (Christian) residents of a locality, as is clear in the discussion of the eruv in Chapter 2 and of an old, formerly white working-class street market in Chapter 3. At their most extreme, exclusion and antagonism towards others who are different are articulated in acts or discourses of racism, sexism and homophobia. Extreme violence, through words or actions, towards others is not the subject of this book, since I am concerned here with the more everyday ways difference is negotiated. It is, however, my strong view that these 'isms' and their effects should be exposed, tackled and addressed wherever they occur. Exclusionary practices, though, operate at many levels, and power is distributed and exercised in uneven and shifting ways. Thus, older people, as we see in Chapter 6, articulate their lack of power in relation to youth, while kids weave connections in the spaces permitted to them by adults. Normative prescriptions as to acceptable social and cultural practices in public – imbued with Western understandings of appropriate behaviour – also operate to define the parameters and inclusions of public space.

There are deliberately many illustrations in the book of urban encounters which are full of delight and enchantment, particularly in the stories of the Hampstead ponds and the Turkish baths. I am interested in exploring different aspects and forms of constraint operating in the public arena which limit the 'coming together of strangers', the 'living with difference', the 'enchanted encounters', the pleasures and displeasures of association and connection. It is only when these constraints and limitations, and the fragile, interstitial and partial forms of connection across difference, are understood that we can begin to think about how to support or construct the kinds of public spaces which may enhance these very connections. As I suggest in Chapter 4 on the Hampstead ponds, one such obstacle is the new culture of risk. The imaginary fears it produces on the one hand, and the official and personal attempts to manage and control it on the other, provide a theme to which many of the chapters allude and return. My hope is that the stories told here (which can be read and dipped into in any order) not only throw new light on encounters with others in public space but also disrupt and expose how normative constructions of public space as accessible to all in fact define and delineate participation, working to include and exclude in shifting, complex and contradictory ways. And, finally, it is hoped that these detailed ethnographies of urban encounters and public space, with the complexities, enchantments and disenchantments entailed, will go some way to destabilise dominant, sometimes simplistic, universalised accounts of public space and help us reimagine urban public space as a site of potentiality, difference and delightful encounters.

2 Symbolic spaces of difference

Contesting the eruv in Barnet, London, and Tenafly, New Jersey

Public space can be material and lived, or symbolic and imagined, though the boundaries between the two are by no means clear. Subjectivities are produced symbolically, discursively and materially through a network of power relations and practices articulated in space in complex and shifting ways. So too urban encounters are woven across spaces that are visible and invisible, performed, experienced and conducted through words and silences, glances and gazes, regard and disregard, acknowledgements and hostilities, all of which are differently embodied. In this chapter, the focus is on an Orthodox Jewish space of connection and encounter, the 'eruv', which, though symbolic, has very powerful material effects. This is a symbolic – and material – space which, though largely invisible, except to those aware of its existence, has in two places in particular – Barnet in North London and Tenafly in the USA – given rise to fierce debate and contestation. These conflicts sharply expose some of the limits of living with difference and normative versions of multiculturalism in the city. Given the eruv's inherently innocuous characteristics – there is no large visible built structure as in the case of the mosque (Naylor and Ryan 2002: 39–59) – it is all the more interesting that the reordering and redefining of the spaces concerned have been so contested.

The eruv has been the object of academic and television documentary interest. Cooper (1998) has explored the relationship between community and space and how, for the opponents of eruvim (the plural of eruv in modern Hebrew), and of the Barnet eruv in particular, the eruv threatens Britain's identity as Anglo-Christian, the British public as a national community based on national liberal values (ibid: 126–127), and the notion of the permissibility of different cultural practices as long as these are performed in private. In an earlier article (1996) she locates this opposition within a commitment by opponents of the eruv to modernist discourse and beliefs, such as the notion of universal public citizenship, existing forms of belonging and the secular character of the public sphere. Vincent and Warf (2002) similarly have argued that the eruv disrupts dominant conceptions of the city built around secular rationality, while stressing that the eruv centrally poses a question about the willingness of authorities and residents to sanction the city as a site of multiple readings. Valins (2000) has explored the idea of the eruv as a window into the complex realities of the institutional sacred geographies of orthodoxy in the context of the (post)modern, largely secular surrounding world.

Through a comparison of these two cases – one in Britain and one in the USA – this chapter seeks to extend and develop these arguments to make four points. The first is that different 'multiculturalisms' have failed to engage centrally with the symbolic which, I argue, when deployed in a spatial context, has powerful material effects. Hall (2000) usefully identifies six versions of multiculturalism: conservative multiculturalism (assimilation of difference into the traditions and customs of the majority), liberal multiculturalism (tolerating only in private certain particularistic cultural practices), pluralist multiculturalism (according different group rights to different communities), commercial multiculturalism (recognising cultural diversity in the market place), corporate multiculturalism (managing minority differences in the interests of the centre) and critical multiculturalism (which 'foregrounds power, privilege, the hierarchy of oppressions and resistance'). Each of these neglects the symbolic sphere. The second point is that opposition to different forms of cultural practice is articulated through different legal or official discourses in different spatial/social/political contexts: the two cases illustrate the significance of constitutional law as the major arena for such arguments in the USA and planning law as the crucial arena in Britain. My third point is that discourses of opposition which draw on legal or official arguments can often mask a more profound resistance from the dominant culture (in this instance, white Anglo-Saxon Protestant, WASP) to 'otherness' which is little more than thinly veiled racism (in this case, anti-Semitism). I suggest, furthermore, that a recognition and fear of how easily racist responses can be mobilised, and of how tenuous is the acceptance of minority cultural practices, underpin some of the opposition to the eruv expressed by liberal and reform Jewish people themselves. I conclude, following Vincent and Warf (2002), with the argument that the multiplicity of imaginings and meanings attached to different spaces necessitates a more nuanced way of thinking about public space and the city.

First, what exactly is an eruv? The Boston eruv website provides a helpful account (http://home1.gte.net/hefter/eruv/history.htm). For traditional Jews, Sabbath is the day set aside for rest and calm away from the fast pace of weekday life, involving a cessation of labour of various kinds. Various restrictions are laid down in Jewish law, including prohibitions against the carrying of objects from private domains to public domains and vice versa on the Sabbath. These public domains include streets, thoroughfares, open areas, highways and so on. Private domains are homes and flats in residential areas which are enclosed and surrounded by a wall and thus closed off from the public areas. In these private areas carrying objects on the Sabbath is permitted.

The purpose of the eruv – which in Hebrew means mixing or joining together – is to integrate a number of private and public properties into one larger private domain; or to put this another way, to redefine the activities permitted in semi-public (or karmelite) space for the purposes of the Sabbath in order that activities normally allowed only in the private domain can be performed (Valins 2000: 579). This is a process of temporal and spatial reordering. Once an eruv is constructed, individuals within the designated area are permitted to carry and move objects across what was hitherto a private–public boundary. This may include anything from the carrying of house keys, handbags (as long as they contain no money) or walking sticks, to the pushing of a pushchair or wheelchair. The construction of an eruv is thus of particular relevance to women with small children and people who are frail or disabled – those effectively excluded by age, gender or infirmity from the public space on the Sabbath.

The practice of demarcating an eruv has been used by Orthodox Jews for 2,000 years and is based on principles derived from the Torah, developed in the Talmud and codified in Jewish Law. According to Talmudic law, an eruv has a very precise definition (http://home1.gte.net/hefter/eruv/history.htm). For an area to be reconstituted as a private domain it must cover a minimum of 12 square feet and be demarcated from its surroundings by a wall or boundary of some sort or by virtue of its topography. Already existing boundaries such as fences, rivers or railways or even rows of houses can serve as the basis for an eruv, but where the boundary is not continuous – broken for instance by a highway – a boundary line must be constructed in order to maintain the enclosed space. The concept here is that where a door separates two rooms in a house, the remaining structure on either side is still a wall, even if there are many openings. The eruv in the modern city is thus the limit case where the notional wall contains many openings with very little solid wall remaining. To construct the enclosure, there are clear vertical and horizontal elements which make up its parameters. To make acceptable the door/lintel combination, an eruv can use existing poles in the street – such as telephone/electricity/cable – or new poles can be constructed. These are joined either by existing wires (usually the lowest in place) or by new wire – as in the case of the Barnet eruv – nylon fishing line or plastic cable. For the pseudo-door to be acceptable the lintel (wire) must rest above the door posts, which can be made by attaching a thin vertical rod to the existing pole to serve as the door post surrogate, or they can act as poles in their own right. In Hebrew these are *lechi*. In the construction of eruvim in the UK and the USA it is these almost invisible objects which have become the site of contestation even though their visibility or intrusion on the street landscape is minimal.

There are eruvim in many urban areas across the globe including Canada (Toronto where it has existed for over 60 years), Australia (Sydney, where the boundary is created from cliff faces, a golf course and fences along Bondi beach, and Melbourne), Belgium (Antwerp), France (Strasbourg), Italy (Venice), South Africa (Johannesburg), and many in the USA. The Barnet proposal represented the largest of its kind in the UK. Eruvim vary in size from a small front yard of a single household, to a large building such as a hospital (allowing Orthodox Jewish medical staff to work on the Sabbath), to matching the boundaries of whole cities as is often the case in Israel. Even the White House is included in the boundaries of the Washington eruv (Vincent and Warf 2002: 35–36). Usually eruvim are distinct spatial entities, although where they overlap due to different communities' non-recognition of each other's eruv, coloured ribbons are attached to the wires to avoid confusion for members of synagogue communities. Ironically, though themselves drawing on an ancient concept, some eruvim are highly modern in their use of the Internet, keeping their members informed of the state of their local eruv

Plate 2.1 Eruv poles barely visible on the left-hand side of this house in North London.

through websites. Typically eruvim are patrolled the day before the Sabbath to ensure that the enclosure is intact and wires are not broken, since they cease to function once a gap has emerged, often indicated on websites by a kind of traffic light system of green and red lights. This eruv boundary is unlike other boundaries in that when it ruptures nowhere inside is safe or unaffected.

For an eruv to become operational in the first instance a civil figure with jurisdiction over the prescribed area has to give permission for it (for which a nominal fee is paid). At one level this requirement necessarily constitutes the group requesting the eruv as dependent and powerless in relation to a state or other official. For example, one Orthodox woman described how an Oxford quad was defined as an eruv for one evening for the purposes of a party on the Sabbath, for which permission was requested from the college rector. In the USA they are generally established by means of a ceremonial proclamation 'renting' the area, issued by municipal authorities; this has been the practice in such cities as Washington DC (including in its boundaries the US Supreme Court), Baltimore, Cincinnati, Charleston and Jacksonville (Becket Fund 2002).

According to two rabbis interviewed in Barnet many eruvim have been established with minimal local objection, often barely entering the consciousness of many local residents. However, in various instances the construction of an eruv has been hotly contested, sometimes over many years. Tenafly in New Jersey, USA, and Barnet in North London are two such cases where opposition was intense but was articulated through different arenas of the state.

CONSTITUTION MATTERS: THE CASE OF THE TENAFLY, NEW JERSEY, ERUV

The Bill of Rights of the United States constitution disallows the state from interfering with the free practice of religion. It was on these grounds that the eruv case in Tenafly was fought (see the proceedings of the United States Court of Appeals for the Third Circuit (No. 01–3301) Tenafly Eruv Association Inc., against the Borough of Tenafly, and the Becket Fund for Religious Liberty (http://www.becketfund.org/litigate.Tenaflly.html)). The controversy began with a meeting between eruv supporters and Mayor Ann Moscovitz in June 1999, where the Mayor raised no concerns about any possible violations of local ordinances, but reported that she had no authority to permit the eruv and suggested a formal proposal be made to the Tenafly Borough Council. At a July council work session, she spoke in favour of the eruv, but met with vehement opposition from some residents who declared that Orthodox Jews would take over the community. One council member voiced his serious concern that 'Ultra orthodox Jews might stone cars that drive down the streets on the Sabbath'. An ordinance of the borough (Ordinance 691 Article V111 (7) (1954)), which encompasses 4.4 square miles, with a population of 13,806, states: 'No person shall place any sign or advertisement, or other matter upon any pole, tree, curbstone, sidewalk, or elsewhere, in any public street or public place, excepting such as may be authorized by this or any other ordinance of the Borough.' Nevertheless, borough officials had often made exceptions: house number signs were visible on utility poles, local churches were tacitly allowed to post directional signs, the local Chamber of Commerce was allowed to affix holiday displays during the Christmas season, and personal private postings like lost dog signs were left unchallenged. Even more politically imbued fixtures like orange ribbons marking a local school's demise due to school regionalisation were ignored by the borough during the protracted controversy. However, without any reference made to the ordinance, the council informally took no further action.

One month later the Tenafly Eruv Association (TEA) approached a Bergen County executive to issue the required ceremonial proclamation, which he subsequently performed in December of that year. Armed with the proclamation, the TEA approached Bell Atlantic (later renamed Verizon) for permission to hang the lechis on the company's poles. No wire was needed since the cable wire already in place provided the necessary perimeter. Permission was granted and with help from Cablevision trucks and crews, the eruv was completed in September 2000. Hearing of the action, the Borough Administrator wrote to Cablevision demanding it be taken down. The action was quickly contested by the Eruv Association which persuaded the borough to allow it 30 days' grace, during which time a formal request was made. By the beginning of November no borough official had mentioned the relevance of Ordinance 691 to the dispute.

Hearings were held on 28 November, and at a December meeting, when Ordinance 691 was brought to members' attention for the first time, the council voted 5–0 to deny the application, and to order Cablevision to take off the lechis as soon as possible. The TEA responded by suing in the District Court, alleging violations of the First (the Free Speech clause) and Fourteenth (the Free Exercise clause) Amendments and the Fair Housing Act (FHA), and seeking an injunction barring the borough from interfering with the eruv. Though recognising that the act of affixing lechis constituted 'symbolic speech', the court concluded that the borough's application of the ordinance did not discriminate against the plaintiffs' religious viewpoint. It similarly rejected the claim that the borough had violated the Free Exercise clause and disagreed that the objective effect of the decision was to discriminate against religiously motivated activity, since the lechis had been ordered to come down pursuant to Ordinance 691 which was a 'pre-existing, neutral law of general applicability'.

Nevertheless the court determined that the council members' improper subjective motivations necessitated strict scrutiny, since they had been influenced by the 'constitutionally impermissible' fear that the eruv would facilitate the formation of an insular Orthodox Jewish 'community within a community'. Nor did the court support the plaintiffs' claim (under the FHA) that they were subject to a discriminatory housing practice, since though the lack of an eruv made living in the borough less desirable it did not result in the unavailability of housing. Injunctive relief was thus not deemed appropriate. The decision was immediately taken to the US Court of Appeals for the Third Circuit, which heard oral arguments on 21 March 2002. In the meantime the Becket Fund filed an *amicus curiae* brief with the Appeals Court, asking for the decision to be reconsidered and for an injunction against the removal of the lechi pending the appeal. Three months later a second amicus brief was filed explaining why the borough's action constituted 'viewpoint discrimination unsupported by a compelling government interest' (Becket Fund 2002).

Under the Free Speech clause constitutional protection is afforded not only to speaking and writing but also to non-verbal acts of communication including expressive conduct or symbolic speech. Thus the affixing of lechis could only be protected under this clause if it could be shown to constitute expressive conduct. The Court of Appeals relied on two precedents in the case – the *Church of the Lukumi Babalu* v. *City of Hialeah* (1993) and the *Fraternal Order of Police* v. *City of Newark* (1999). Striking in the report is how the context gives meaning to the symbol: for example, the burning of the American flag in protest against the Reagan administration's policies at a 1984 Republican Party Convention was deemed expressive due to its overtly political, intentional and apparent nature, as was the attachment of a peace symbol to the American flag in protest against the invasion of Cambodia and the killing of student demonstrators at Kent State only days before the protester's arrest. In contrast, in the eruv case the court concluded that the Association had not met the burden of showing that affixing lechis to poles was 'sufficiently imbued with elements of communication' to be deemed expressive conduct. Rather, they deemed the eruv served purely functional, non-communicative purposes indistinguishable from those of a fence surrounding a yard or a wall surrounding a building since no inherent message – ideological or otherwise – could be discerned from the lechis themselves: they simply delineated an area where certain activities are permitted. The Free Speech claim thus failed.

The Free Exercise clause, which binds the borough to the Fourteenth Amendment whereby 'Congress . . . shall make no law . . . prohibiting the free exercise [of religion]', was likewise scrutinised. Under this clause, if the law is generally applicable and burdens religious conduct incidentally, then no protection is offered. If the law is not neutral or is not generally applicable, restriction of religious conduct violates the Free Exercise clause. The clause's mandate of neutrality towards religion also prohibits government from 'deciding that secular motivations are more important than religious motivations'. In one case (*Church of the Babalu* v. *City of Hialeah* (1993) *Lukumi* 508 U.S. at 537), the Supreme Court thus invalidated an ordinance against the unnecessary killing of animals which was used by local and state officials to ban animal sacrifices during Santeria religious ceremonies, while at the same time exempting secular activities such as hunting or slaughtering animals for food. A city's police department's no-beard policy was also cited as coming under the Court's scrutiny (*Fraternal Order of Police* v. *City of Newark* (1999) 170 F.3d at 364–66) when it was found that medical exemptions were allowed while exemptions had been denied to two Sunni Muslim officers whose faith required them to grow beards. In the eruv case the borough asserted that strict scrutiny by the court should not apply on a number of legal grounds, which included the argument that the plaintiffs had not shown that the removal of the eruv would substantially burden their religious practice. A Free Exercise claim was thus not valid since the eruv was an 'optional' practice. The court, however, did not deem it necessary to consider whether the borough's characterisation of the eruv was accurate. Rather they needed to consider whether the borough's invocation of Ordinance 691 – with its objective of avoiding visual clutter and maintaining control over municipal property – against the lechis was persuasive.

The following strict scrutiny of the court highlighted the permission given by the borough to private citizens to affix all sorts of materials to utility poles, from house numbers to lost animal notices. Though the borough rested its case on the more permanent nature of the lechis, the court argued that this was precisely the sort of reasoning that the Lukumi and Fraternal Order of Police decisions forbade. The court concluded that the plaintiffs were not likely to prevail on their Fair Housing Act claim and did not present a viable Free Speech claim, but were reasonably likely to show that the borough had violated the Free Exercise clause by applying the ordinance selectively against conduct motivated by Orthodox Jewish beliefs. On 24 October the Third Circuit issued a decision finding that the District Court should have preliminarily enjoined the borough from removing the lechis pending a trial. It reversed the District Court Judge's denial of injunctive relief and ordered him to issue a preliminary injunction barring the borough from removing the lechis. As an Eruv Association spokesperson, Chaim Brook, put it: 'I'm hoping that they'll conclude that there's really no reason to fight this further. It's a waste of time, money and energy for all of us' (Brauner 2002).

This necessarily detailed account reveals the centrality of constitutional law to the eruv proponents' and opponents' case, in stark contrast with the British case where the key arenas of permission or resistance were those of planning and local government. It also highlights the importance of affect, or more particularly how fears of anti-Semitism underpinned the responses of some of the Jewish people involved. This is revealed further in an article in the *New York Times* which describes Mayor Ann Moscovitz's own arrival in the suburb 39 years earlier, when she was welcomed, but only up to a point. She remembered the real estate agents directing her away from various parts of the affluent Bergen County town:

> 'Well Mrs Moscovitz,' with an emphasis on Moscovitz, 'we can show you houses but no one will sell them to you, so let's not waste your time.' She moved to an area where they accepted her, raised her three children as reform Jews in a growing Jewish community and got involved in civic affairs. Five years ago she was elected mayor. Now she's accused of doing to Orthodox Jews what was once done to her . . . welcoming them, but only to a point.
>
> (Purdy 2001: 35)

According to Purdy's analysis, despite new celebrations of diversity in the once homogenized suburbs, some celebrations don't sit well. Orthodox Jews stick out with their separate religious schools, their Sabbath walks to the synagogue, their distance from regular Saturday shopping or sports and 'their air of separateness – some say superiority – [which] often draws the loudest complaints from other Jews'. From this journalist's point of view, the debate around the eruv was Seinfeldian, ostensibly being about nothing, simply black cable casing on telephone poles which blended easily with the usual black cables. But he also admitted that below the surface there was the question of 'how much any one piece of a diverse town's mosaic should alter the big picture', and more specifically what he described as the family feud between Orthodox and non-Orthodox Jews.

The article quotes various different responses: one from a Holocaust survivor mirrors similar responses in the London eruv: 'They are building their own ghetto'; while Rabbi Mordechai Shain was reputed to have said that the Orthodox irked some of the other Jews because 'when you see people doing it the right way, you feel a little guilty' (Purdy 2001). And as in Barnet, fears were expressed that hordes of Orthodox Jews would move into the area if the eruv went ahead. The Mayor's concern was thus expressed: 'an influx of Orthodox Jews would jeopardize the acceptance and progress the Jewish population in the borough had achieved'. From her point of view, constructing the eruv without permission had caused the problem: 'they have gotten to know us, the old timers in town, and accept us. Now Jews came in and violated the law, did something sneaky, and it's bad for all Jews'. (One wonders if the posting of lost animal signs evokes a similar discourse of sneakiness.) The leader of the Eruv Association, Chaim Brook, who had chosen to move to the borough because he could easily socialise there, described how since the eruv feud he had felt people were staring at him in the supermarket when he was wearing his skullcap. As I discuss more fully in relation to the Barnet case, what is revealed here is the tenuous sense of acceptance in the locality that some Jewish people felt, which the eruv was seen to threaten yet further.

PLANNING AND LOCAL GOVERNMENT: ENGLISH SITES OF CONSERVATISM AND THE BARNET ERUV

The boundary of the Barnet eruv stretches for 11 miles around several wealthy neighbourhoods including parts of Hampstead Heath, a popular, mainly middle-class recreation site, Golders Green and Hampstead Garden Suburb. Most of the eruv is marked by existing boundaries, including two three-lane highways – the M1 and A1 – railway lines and rows of terrace houses. To complete the eruv the construction of 80 poles was required, from which strands of fishing line, 0.3 mm thick, could be strung to complete the enclosure mainly across road junctions, but also at other key points such as in Hampstead Heath. The line was to be approximately 1,000 yards in total and not distinguishable from other wires from street level. Given the presence of tens of thousands of telegraph poles, lamp posts and street signs, the eruv posts at the material visible level represented an insignificant addition to the street furniture.

The idea for the eruv was first mooted in 1987 when an Orthodox rabbi, Rabbi Kimche (interviewed for this research), with the following of his congregation, joined with the United Synagogue, which is the largest grouping of synagogues in Britain, to form the United Synagogue Eruv Committee. Like eruvim around the world, the eruv was seen by its proponents as a harmless vehicle which, by reclassifying a semi-public (karmelite) domain as private for the purposes of the Sabbath, enabled Orthodox Jews excluded by gender, disability or frailty to participate more fully in the community on what is seen as a day of rest and simplicity. As an Orthodox woman interviewed put it:

Figure 2.1 The eruv boundary in Barnet.

Source: www.nwlondoneruv.org

> Obviously it doesn't matter to anyone else, but for us is it is a space where we can go about our business . . . where we can live as a village . . . popping in and out of each other's doors, take a baby to a friend with a baby for the kids to play together or eat in each other's houses which we do a lot . . . and if you are disabled you can't go out on the Sabbath to celebrate a birthday with friends or go to the synagogue.

A planning application for the eruv was lodged with Barnet Council in August 1992, asking permission for the poles and wire. The claim was that over 10,000 people in the borough would benefit. Over the following six months nearly 1,000 letters and several petitions were received by the council, either supporting or objecting to the eruv. The first and marginally modified second applications were rejected on the grounds that the poles and wire were visually obtrusive and constituted unnecessary street furniture which was detrimental to the character and appearance of the street (contrary to various policies outlined in the Unitary Development Plan). The committee was also advised that the council, as highway authority, could not grant consent under Section 178 of the Highways Act 1980, since the proposal to use street lighting columns to support the wire was contrary to the operational practice of the highway authority. After further vociferous correspondence, including the Royal Society for the Protection of Birds' objection to the wire's supposed impact on bird life, an appeal – with further modifications to the application to meet with the objections – was lodged at a Public Inquiry on 30 November 1993. A report from the Controller of Development Services to the Town Planning and Research Committee on 27 October (item no. 4) laid out various conditions if approval were to be granted: that the development be commenced within five years, that the posts be treated with anti-vandalism paint above 2 metres (to safeguard the security of adjacent properties), and that no trees be affected.

A letter was submitted by the applicants in support of the application, stressing the importance of the eruv to a substantial religious minority, and the minimal impact of the proposal on the locality. The public consultation unleashed strong opposition as well as continuing support. One of the most vociferous opposition groups was the Hampstead Garden Suburb Conservation Area Advisory Committee. Hampstead Garden Suburb was built in the early twentieth century according to the principles of Dame Henrietta Barnett, a cosmetics heiress turned social worker, and Raymond Unwin, an architect planner/social reformer. It is seen as a unique site and many local residents express strong commitment to its original ethos. To quote its website (http://www.hgs.org.uk/history/h00000000.html), it was 'socially, politically and spiritually a new kind of creation: a joint co-operative endeavour by a group of like-minded citizens'. The original architectural style and its conservation are integral to the project, as are the local societies such as the drama society whose summer performance during the conflict year was (a rather apposite) *Thomas Becket*.

Of the five petitions, three were against the eruv's construction. Although there were objections from non-Jewish residents, much of the opposition came from Jewish residents and some rabbis, as in Tenafly. The petition from 42 residents of the suburb focused on the 'unacceptable visual blight' and on the view that the proposal 'would cause serious disturbance to social harmony in the area. Deep offence and hostility would ensue if religious hardware of this kind were erected on public roads and open spaces.' In the 49 letters received (including one joint letter from eleven residents), there was a strong nostalgia for the past, with the early principles of the suburb and Dame Henrietta Barnett's original vision being cited. Much of the conflict ostensibly revolved around the visual questions and conservation, since the area is strictly controlled by planning legislation which requires that even new door or window frames are constructed in the original style and changes to paint colours are also restricted. Notwithstanding this, one of the key opponents to the eruv, Lord McGregor, had enlarged his own house's windows, a fact rather humorously pointed out and challenged by the Chair of the Eruv Association, Edward Black, in a television documentary on the eruv.

Plate 2.2 Hampstead Garden Suburb – quintessentially English.

Of the 806 letters of representation, 188, including one with twenty signatories and a second with eighteen, objected to the application. The letters in support are not of interest to our argument here, ranging as they did from comments that there would be no discernible impact on the community to the assertion that there were many other eruvim in major cities in the world which appeared to have no detrimental effects. There were many reasons for objection, ranging from a stress on the supposed visual intrusion to more emotive and affect-laden comments. These included the view that the delineation of a territorial boundary created a provocative and overpowering situation for residents, that a ghetto would be created, that it represented an invasion of space for non-Orthodox Jews and Gentiles alike, that it would attract vandalism and unnecessary racial hatred, that it was reminiscent of Nazi Germany, that it would cause anti-Semitism, that it meant the withdrawal of the freedom of movement on public land, that it would attract inter-racial and inter-religious problems (particularly a split within the Jewish community), that the permanent installation of private religious symbols on public highways was a historical anathema, and that it was a violation of the principle of democratic government. Thus for example the occupants of Wildwood Road (London Borough of Barnet 1993: 46) felt that the poles were amongst other things 'a hazard to the wheel chaired and poorly sighted, were pandering to the needs of a minority group for 1 day in each week, were an eyesore, would attract graffiti and vandalism'. There were also the more property-related concerns (capital value fears) – that the eruv had the potential to change local demographics, reflecting the notion that once houses were advertised as being located within the eruv more Orthodox Jews would be attracted into the area. Yet behind this lay the idea of an invasion, which for Rabbi Kimche was appalling: 'I didn't know there was a quota on Jews, imagine saying that about blacks – it's offensive.'

Some of the joint letters from 34 local residents of 11 November 1992 and 4 February 1993 raised further legal and other issues. The first recognised that there were no strict planning guidelines given the unprecedented nature of the eruv, and no relevant guidance other than the statute and decided cases which themselves drew on earlier cases which had a bearing on the eruv proposal. Parts of this letter raised some interesting issues (London Borough of Barnet 1993: 104). A case is cited which decided that private interests were a material consideration (*Stringer* v. *Min of Housing* [1970] 1 WLR 1281):

> We considered the private interests of individuals who have to face the permanent prospect of having (within one foot of the front boundary wall of their house) a pole supported by wires/cables, in aid not of a public service (light/telephone etc), but rather in aid of a sectional interest with which they may not identify.

And later:

> 'The purpose of the eruv is to convert the entire area within the boundary into one, notional jointly owned domain.' This quotation is from the proposal. If this conversion were notional only, there would be no difficulty. Unfortunately for Barnet's freeholders and road users the conversion although primarily symbolic is effected by real posts and wires upon this most public of amenities, the high way . . . in this case, you are driven to ask whether visual religious symbols having a particular meaning for some, are 'appropriate' to the public highway, which historically contains only installations of a strictly functional nature, designed and located to provide a public service. This symbol means different things to different people: such is the nature of this kind of symbol. To A the eruv symbolises the 'Sabbath limits'. To B 'a desecration of the Sabbath' (a group of Orthodox Rabbis have so proclaimed), to C 'a communal dividing line' to E 'the walls and gates of the ghetto', to F 'non-functional street clutter' to G 'a focus for anti-Semitism' and so on.

In the letter of 4 February from another such group we find a similar set of arguments (London Borough of Barnet 1993: 114): 'objects of this kind, having such multivalent significance to the public are not appropriate to the public highway'.

The application for planning permission having failed, the case went to appeal at the Department of the Environment Court in 1994 which concluded on the basis of the evidence before it that there was no good reason to disqualify the eruv, with the result that Barnet was required to grant permission. From the Court Inspector's point of view: 'however hard the objectors say they are not prejudiced they are wrong: they are prejudiced on moral, social or religious grounds' (quoted in Vincent and Warf 2002: 44). For the following eight years all sorts of planning, legal and other devices were deployed to delay the eruv's construction, with continuing objections from the opponents of the eruv and stalling by Barnet Council (*Hampstead and Highgate Express* (hereafter *Ham and High*), 26 June 1998, p. 5; 2 October 1998, p. 11; 23 October 1998, p. 11; 25 June 1999, p. 15; 13 August 1999, p. 9). Even the Superintendent of Hampstead Heath warned that the poles and wires at Wildwood Road would seriously damage nearby trees on the heath (Gilbraith 1998: 5). In my own interview with him he described the conflict as the single most divisive issue he had encountered. Rabbi Kimche described one problem with the siting of the poles as the need to apply to five different authorities for permission: 'the Water Board would request it be moved 2 feet one way and the cable TV company 2 feet the other way'. The colour of the poles became similarly contentious: the Eruv Committee proposed sage green only to find that the council's Public Works Committee had decided that the colour should be subject to consultation, using this to further delay the scheme.

Permission to construct the eruv was finally granted in the summer of 2002. Writing as the plans were finalised Steven Morris quotes one of the members of the eruv boundary opponents (10 August 2002): 'We feel that our human rights will be affected. It's a monstrous thing, an affront to civil rights'. Part of the problem appeared to be that people felt they had no choice about their own property being used to mark the line. A letter sent to the Chairman of the United Synagogue Eruv Committee on 25 October by a number of signatories at a house in the area reiterates the concerns motivating the objectors:

> We strongly object to the poles being erected so close to our homes because they interfere with the enjoyment of our properties. We acknowledge that the poles have religious significance. However it is the religious significance and the fact that the poles are so close to our homes that lie at the root of our objections. We will constantly be reminded of this religious significance every time we arrive at and leave our homes. We consider this to be an unfair and unreasonable imposition of a religious belief upon us. It also amounts to an unacceptable public pronouncement of religious belief of others upon our private and family lives. . . . This interference also amounts to disrespect for the privacy of our homes.

Also in the letter we find the claim that these residents have been told by estate agents that the eruv will make it harder for them to sell their homes and cause a consequent reduction in their value (elsewhere it is claimed that house prices will rise as a result of an Orthodox Jewish influx). The Chairman of the United Synagogue Eruv Committee sees a tautology here: 'to say they acknowledge (actually assert) the posts have religious significance is like saying the road to a synagogue has religious significance' (personal interview, 5 November 2002). But this is to overlook the power of the symbolic.

To establish some control over the discursive field, the Eruv Committee, through the Chief Rabbi's press office (press.office@chiefrabbi.org), posted a list of supporters of the eruv who were available for interview. The list reveals the importance of the eruv for women. Jodi Benjamin (33) was an expecting mother and head of an international new media company, who felt that being able to build relationships on the Sabbath is crucial to any family. Anushka Levey (28) is described as a hard-working barrister and mother expecting her second child, who had felt isolated on the Sabbath since her daughter was born – imprisoned in her own home and unable to join her husband in religious and social activities as she once did. Millie is a disabled full-time mother who is wheelchair bound and has been stuck in her house on the Sabbath for many years, even unable to attend her own sons' bar mitzvahs. Unable to attend the synagogue, she feels outside the community. These are three of the eight named profiles telling similar stories of exclusion and imprisonment through the need to care for others or through disability restricting movement on the Sabbath.

Discourses of dissent

The history of these two cases has clearly established part of my first argument – that the symbolic has distinct spatially configured material effects which are not incorporated into the dominant versions of multiculturalism in either the UK or the USA. We have also seen how the sites of opposition and resolution to the eruv differ very significantly between the two countries – in the USA constitutional law is deployed, in Britain, planning and related arguments represent the key site. Naylor and Ryan (2002: 55) found a similar politicisation of planning in their study of London's first mosque. Another difference between the two countries was the relationship between church and state. In both cases the legal concern was whether the religious character of the eruv had been recognised and treated as a consideration (positively or negatively). But in Britain the religious character of the eruv was considered unimportant in the policy arena since only the physical structure was deemed to matter. However, in the USA the state must be neutral between religions, since the constitution is founded on the separation of the church and state, and a guarantee of religious freedom is enshrined in constitutional law. This is the mythos of America, which may explain, in part, the success of Orthodox Jews in establishing eruvim in many American cities. In the Tenafly case, as we saw, much of the argument rested on whether a refusal of the eruv constituted a breach of the Fourteenth Amendment – the Free Exercise clause. In contrast, in the UK the church and state are intertwined, the Queen bring the head of the Church of England – a thoroughly Christian institution – and bishops being granted a seat in the House of Lords by right. In Britain minority religious practices are seen as a private matter to be conducted in the private domain of the home. Marx saw the problem this produced for the Jew (and of course, any non-Christian religious group) (2003: 17) as irresolvable:

> In demanding his emancipation from the Christian state he asks the Christian state to abandon its *religious* prejudice. But does the Jew give up *his* religious

> prejudice? Has he then the right to insist that someone else should foreswear his religion? The Christian state, by its very nature, is incapable of emancipating the Jew. But . . . the Jew, by his very nature, cannot be emancipated. As long as the state remains Christian, as long as the Jew remains a Jew, they are equally incapable, the one of conferring emancipation, the other of receiving it.

The Hampstead Garden Suburb drama club's performance of *Thomas Becket* highlighted this all too clearly.

Reflecting on resistances to the eruv, I want to concentrate more particularly on the Barnet issue to consider the question of resistance to othered cultural practices. As discussed earlier, a large part of the Barnet eruv is located in the deeply English space of the garden suburb, imbued with English notions of architecture, landscaped gardens, country village living, and community with all its connotations of homogeneity and conforming. What is clear is that the eruv centrally threatened these ideas at their core. In a film made for Channel 4 on the Garden Suburb in the early 1990s, shots of the rehearsals for *Thomas Becket* are cut with an interview with a very upper-class-sounding Lord McGregor and his wife talking of these 'eruvites' almost as an alien species. Another woman interviewee also stressed a notion of history and Englishness (not Britishness):

> I'm not ashamed of being against it – it's a reasonable thing. I do research on the sixteenth century – such a business it was then with the Catholics and the Protestants. It's important to keep England stable – do you know what I mean? They are a minority race and they should be reasonable about it. All this business now about race – I'm sympathetic about it – but we are a small island and in order to keep it stable everyone should think about everyone else – otherwise there will be all sorts of rows and we won't be England any more. Even my neighbour who is Jewish felt the same. . . . We don't have those things in England, we should have freedom and not have to have something imposed on you. If they want that they can go to Israel if they feel that strongly. It is an imposition on my freedom.

Whose freedom and whose England? we may ask.

A paragraph in the Town Planning Report resisting the proposed boundary crossing a main highway revealed a similar defence of English cultural practices:

> It seems unfortunate that the committed local interests involved were not informed at an early stage that the law of the highway embodies fundamental English ideas about the liberty of the subject. It speaks of 'the rights of the public' 'all Her Majesty's subjects' 'the public generally' 'every member of the public' and specifically that 'there can be no dedication of the highway to a limited section of the Public'. The Rabbinnic legal concepts of the Eruv are out of keeping with the laws of England, which in the case of the basic law affecting the public highway, derive from time immemorial. As a matter

> of law, those members of the public, who have sensed that their legal and democratic rights are threatened by the pursuit of this scheme are right.

Such claims did not emerge in the Tenafly debate. The multicultural settlement in the USA, though not uncomplex or unfraught, seems nevertheless to involve a more heterogeneous notion of what it means to be American. Rabbi Kimche would agree (interview October 1999): 'America is a multi-ethnic society, England isn't – English established with a capital E – is Christian – and the Queen is head of the church – it is seen as corrupted by other religions.'

Much of the discourse of opposition was also framed in terms of Enlightenment notions of rationality and modern Western understandings of reason, progress and order. These notions coded here in planning law, resting as it does on a utilitarian premise of satisfying the needs of the greatest number of people, come into conflict with an entirely different system of thought. The eruv is a spatial concept which draws on a pre-modern system of rationality derived from a different legal framework which drew its precepts from an ancient religious text. This clash between the symbolic city and the modernist rational city established within a legal/technical framework draws on cultural practices which cannot easily be recast in other ways. An Orthodox woman interviewed put it this way: 'our legal system can't function if their legal system doesn't sanction it – they can't understand that – either it's separate or it isn't . . . it's a cultural thing not a physical thing.' As a result, many of its opponents saw the eruv as an irrational, even crazy, idea which needed to be resisted on this basis.

The eruv was seen as threatening in another way, in that it challenged and disrupted conventional notions of the public–private boundary underpinning modern liberal thought. For the public realm to be the space of the universal disembodied citizen, difference must remain privatised. Feminists for a long time have emphasised the gendered as well as symbolic nature of this divide (Pateman 1988) which has consigned women to an undervalued private sphere and precluded their full participation in the public realm of work, citizenship and politics. They have also drawn attention to the fluid nature of this boundary where home and work, production and reproduction, are mutually interdependent and constitutive. Nevertheless, the idea that difference is allowable in private rather than in public, that difference is a private matter for expression outside the public gaze, remains thoroughly embedded in notions of the social in Western thought. Cooper (1998: 127) defines this as a cultural contract – an Anglican settlement – where minority practices are acceptable if performed privately and where public space should reflect Anglican values which derive from English history and its Anglo-Christian heritage. The eruv, by redefining the public (or more precisely semi-public) as private for the purposes of the Sabbath, thus raises a major problem since it exposes difference to the public gaze, it allows for the public expression of minority beliefs and cultural practices, even though symbolically, to those involved, this public space in a temporal/spatial sense is now private. Cooper (1996: 537) points out an interesting contradiction here. While the eruv allows Jews more freedom to be in public on the Sabbath, 'at the same time, an eruv normalizes orthodox Judaism by enabling observant Jews to come out as "ordinary citizens". Yet, the essential otherness of the orthodox Jew remains. The eruv's danger is it allows such "otherness" to find public expression.'

Given then the resistance from the dominant majority to cultural practices seen as other or even alien, I was interested to explore the question of why in both cases there was also vociferous opposition from Jewish people living in these communities. This came from two quarters – other Orthodox Jews and more acculturated or secular Jews. Objections from the former group were easier to grasp. Orthodox Jews saw the initiative to create the eruv as contravening Talmudic law and the sanctity of the Sabbath by allowing people to move more easily and enabling a greater mixing of the sexes, thus encouraging infractions of the law (Valins 2000: 583). As Valins points out there is a destabilisation here of temporal divisions between (sacred) Sabbath and (profane) weekdays through the spatial boundaries of the eruv (ibid). As far as non-religious Jewish, or liberal Jewish, hostility is concerned, this seemed to derive from a different source – discomfort at being publicly exposed as Jewish and as a result threatening a precarious sense of acculturation. Here I want to suggest that the success of multiculturalism in the UK in particular, but also in the more 'WASP' parts of the USA, resides in the adoption by minority groups of Anglo-Christian norms and cultural practices. This, it could be argued, is a central tenet for the white majority to subscribe to multiculturalism. As Hage (1998: 53), writing in the context of Australia, suggests, 'a national subject born to the dominant culture who has accumulated national capital in the form of the dominant linguistic, physical and cultural dispositions' will aspire to the fantasy of an imagined space – the 'White Nation' – in which s/he can legitimately control and spatially manage others. For minorities acculturation can often be a precarious state of affairs, dependent, in part, on conforming to normative ideas of acceptable behaviour.

Writing in the German context, Robertson (1999: 286) suggests that

> acculturation without integration made it often difficult for Jews to avoid feeling uneasy in the company of non-Jews. Sooner or later one would be reminded that one was a Jew and that, in many people's eyes, this was incompatible with being a German. And the individual's experiences were overshadowed by the assumptions of the 'emancipation contract', whereby Jews were accepted into German society on condition that they behaved themselves properly – that is, they did not behave in an obviously 'Jewish way'. Some conspicuous reactions took two linked forms. One could maintain the assimilationist stance and try to become more German than the Germans; or one could internalise anti-Semitic hostility.

In attempting to explain this phenomenon Robertson draws on a paper by Kurt Lewin written in 1941 (Robertson 1999: 287–288) which suggested that some members of underprivileged groups who feel ashamed of that membership tend to feel anxious to join the larger community outside but are aware that such a move is constrained by the fact that the community identifies them with the underprivileged group. In Lewin's view, logically this could lead to aggression towards the wider community. However, such an attitude could be experienced as too dangerous since the majority is perceived as too desirable and too powerful to be attacked, one result therefore being an identification with the aggressor – the majority excluding community.

This is not to say that there are not all sorts of rational and affective reasons for the resistance to the eruv expressed by non-Orthodox Jewish people in the areas claimed – public exposure may well, for instance, open up a space for anti-Semitism and aggression from racist or fascist elements. One of the objections that recurred frequently in the interviews and written reports in Tenafly and Barnet was the claim that the eruv, as a powerful symbolic space, produces an idea of segregation which has a strong historical resonance in the space of the ghetto. As one group in Barnet put it:

> as Jewish refugees we arrived in England from Hitler's Europe. Every time we leave our house we will be reminded of the concentration camps and the division of areas for Jews and non-Jews alike . . . our sister survived Auschwitz and we are worried of her reaction when confronted with these posts and wires.
>
> (London Borough of Brent 1993: 22)

Evident here is the fact that the competing meanings of the eruv differ starkly and that these meanings have symbolic power. A space that for one person may simply represent a facilitative device which means nothing to outsiders for others is associated with the idea of a concentration camp and marked as a horrific space. This is a complex arena since once a space has such negative connotations for some people, it is not clear how, or if, these meanings can be shifted. Yet in both Tenafly and Barnet, the evidence was that for some of the opponents the eruv threatened a sense of hard-won acceptance in a hostile country, which, if we recall the estate agents' comments to Mayor Moscovitz on her arrival in Tenafly some 40 years earlier, is no surprise. Thus for Jewish people, consciously or unconsciously in this context, opposition to the eruv could be seen as a form of self-protection against the potential exclusion and racism of the dominant culture.

CONCLUDING REFLECTIONS

The eruv represents an unusual urban boundary in a number of ways. Though it resembles other boundaries in the city – postal areas, school catchments and so on – in its virtual invisibility it is different. What brings the eruv into being is a series of rituals, a performativity where new identities, spaces, social practices and notions of the private are constituted. It is not simply a question of the construction of an eruv, rather it is the routinised and repetitive recognition of the boundary by its users, and the vigilant maintenance required to keep it intact, that maintain it and keep it alive. It is this perhaps that makes the eruv such a potent space, and explains why those opponents whose houses formed part of the eruv boundary were so vociferous in their objection to it. Yet from some of the rabbis' viewpoints, the symbolic power of the boundary for those in opposition was a surprise. As Rabbi Kimche put it:

> if you ask people in Baltimore 'do you have an eruv?' . . . they wouldn't know. People say you have no right to use my road as a boundary but if the GPO say your road is a boundary for the postcode . . . they wouldn't care. They say you're making this road a perimeter of Jewish people, but it isn't like that, we're putting together a very simple and innocent and almost invisible facility. If it means something to us and not to you why should you care?

The reform Jewish Rabbi Geoffrey Newman made a similar point. 'it's only of interest to those who care – that's almost magical . . . it has a great deal of meaning for those for whom it matters. . . . It's an inner meaning projected onto an outer reality . . . I'm totally in favour of it . . . it just doesn't have much meaning for me but I'll defend them to have the right to do it'. Such a defence presumes a voluntarism where symbols only have significance for those who choose to recognise them.

But my argument here is that the power of symbols in this conflict over space is precisely the point. The fact that legal or planning approval has to be granted is, in some sense, to submit to the eruv. Or to put this another way, the very fact of protesting and opposing the eruv gives it legitimacy, meaning and efficacy. For its opponents, the eruv 'stains' the wider space within which it operates (Cooper 1996: 534) while potentially having very real effects on local house prices (though how much this claim can be substantiated in the British context is open to question). Nevertheless, it was a concern which was consistently expressed in both cases. For example in the Town Planning Report from Barnet we find: 'there are also many immediately adjoining houses, one of which is planned as being within the Eruv and one of which is without. The one within may attract a higher price, from some applicants, and this could even complicate banding for council tax' (London Borough of Brent 1993: 104).

The eruv thus raises the question of whether competing meanings of space can, or should, be resolved in the policy arena and whether policy initiatives can confront the symbolic. In this instance the eruv held both positive and negative connotations at the same time, and this was a central problem. There is no easy answer here, but if the symbolic power of space is to be taken seriously, particularly in the context of competing cross-cultural claims, policy responses will need to confront these rather thorny questions raised. One such issue is that minority groups requesting permission from the state – in whatever form this takes – are differently (and less powerfully) positioned from those imbued in the dominant culture, of which the state may be imagined to be an integral part. In the eruv case, the opponents certainly articulated a more confident relation to the various arenas of the state with which they came into contact.

This recognition of the symbolic power of space has thus also to be integrated into the wider arguments around equality and difference. This does not mean acculturation to the dominant culture, leaving difference to be tolerated in the private sphere. It means difference, symbolic or otherwise, finding expression in both private and public spheres. This is the politics of recognition for which Nancy Fraser (1998) argues, where people's 'heritage is encouraged, not contemptuously expected to wither away' (Modood 2003). Such a challenge to earlier liberal formulations of equality in the public sphere, which were predicated on the privatisation of difference, involves a widening of the national culture to include the symbolic worlds of minority cultures even if this is uncomfortable and destabilising to dominant norms. Dismissing the eruv as crazy or irrational is in effect to deny this minority group of residents in the city equal citizenship embedded in the recognition of difference. Instead the multiplicity of meanings attached to different spaces necessitates a much more nuanced and innovative way of thinking about public space and the city.

POSTSCRIPT

Five months after the Barnet eruv had started operating, a 60-year-old man received a 60-hour community punishment order for sabotaging the eruv wire with a metal instrument. He was quoted as saying: 'I don't know what all the fuss is about. It's only a stupid cable. They shouldn't be in the f***ing country anyway' (*Ham and High*, 3 October 2003, p. 1). This struck me as a salutary reminder of how tenuous the multicultural settlement in cities can be.

3 Nostalgia at work

Living with difference in a London street market

For the urban *flâneuse* one of the great joys of the city and its public spaces is the local street market with its ramshackle stalls, intriguing smells, banter and chatter, surprise bargains and cosmopolitan crowds. Entered for the first time, a market is easily seen as the quintessential space of connection, interconnection, easy sociality and living with difference. Such a description rings true for markets in many cities of the world and in London particularly. Yet beneath the surface, in the textured and lived worlds of the traders and locals who frequent these spaces there are many stories to be told of a street market – some of vibrancy and laughter, some rather darker and more ambivalent. In this chapter I look at one such market, Princess Street, in an inner London borough. In many respects this is a market where there is co-mingling between people of different ethnicities and races, young and old, with and without children. There is a story to be told of this market as a positive space of urban encounter – which, in many respects it is, particularly at the level of casual everyday life, where people chatting in the streets and shops, or standing around outside with a cup of coffee or a beer, are a common sight. In this it is like many other street markets in towns and cities in the UK which to a greater or lesser extent can be described as key sites of sociality and engagement (Watson and Studdert 2006).

This is not, however, the thrust of the narrative presented here. Instead I focus on the complexities of class, race and ethnicity and their intersections in a locality overwhelmed by the changes wrought by the global/local social and economic shifts of the last few decades. What I aim to show is that though markets can be places of magical and serendipitous encounters in public space, such connections and sociality can be undermined if the area in which a market is situated is neglected by local authorities and others involved in the production and management of this crucial public space. In other words, in places that have been forgotten, lacking leadership or stewardship, and are neglected by key players with public or private resources to allocate (Markusen 2004), markets are less likely to thrive. In particular I argue that in Princess Street Market encounters between the longer-established British citizens – including black British – and the newcomers to the neighbourhood from many different parts of the globe, many of whom came as asylum seekers, are mediated through discourses of nostalgia, precisely because the market has been subject to neglect for some years. These discourses act to

Plate 3.1 Everyday life in a London street market: Ridley Road, Dalston.

shore up the notion of an imagined happier cosmopolitan neighbourhood in the past, and to interrupt and militate against the possibilities for an easier co-mingling of difference in the present.

PRINCESS STREET MARKET

In *Second-class Citizen* (first published in 1974) Buchi Emecheta tells the tale of a young Nigerian woman who from eight years old had a dream – 'the sort of dream which originated from nowhere, yet one was always aware of its existence. One could feel it, one could be directed by it; unconsciously at first, until it became a reality, a Presence' (1987: 7). Adah's dream was to become a 'been to' – a Lagos phrase for those who had been to England (ibid: 29). The story she tells is one of immigration, race and the city during the 1960s. The location she chooses is Kentish Town in North London. Adah follows her husband to London, he to study accountancy, she to read librarianship, leaving a good job as a librarian back in Lagos.

> Later, in England, writing about that time of her life almost with nostalgia, she used to ask herself why she had not been content with that sort of life, cushioned by the love of her parents-in-law, spoilt by her servants and respected by Francis's younger sisters.
>
> (ibid: 28)

This book is a telling place to start an exploration of the role of nostalgia in Princess Street Market today, an area with a history not unlike the one that Emecheta describes. It is an evocative tale of how her dream was shattered as she encountered the racism, exclusion and segregation of black people in Britain in the post-war years. It is also a tale of an area of Central London before gentrification made much of the housing only affordable to those in the media or the financial services (Hamnett 2000). Adah's attempts to find somewhere for her family to live provide probably the most illustrative moment of the racism which she encounters and for which she is entirely unprepared. Having made an appointment by telephone with a prospective landlady, she sets off for the house in Hawley Street which she describes as being:

> in a tumble down area with most of the surrounding houses in ruins, and others in different stages of demolition. The area had a desolate air like that of an unkempt cemetery . . . Adah did not mind the ruins and demolition because the more insalubrious the place was, the more likely the landlady would be to take blacks.

But Adah's hopes were soon dashed. As the landlady opened the door she:

> clutched at her throat with one hand, her little mouth gasping for air, and her bright kitten-like eyes dilated to their fullest extent. She made several attempts to talk, but no sound came. . . . But she succeeded eventually. Oh yes, she found her voice, from wherever it had gone previously. That voice was telling them now that she was very sorry, the rooms had just gone. . . . Adah could not even utter the pleas she had rehearsed. The shock was one she would never forget.
>
> (ibid: 84–85)

In the end Adah and her family find a house in Willes Road (where in 2003 houses sold for an average price of £600,000) which belonged to a fellow countryman:

> Approaching the street from the Queen's Crescent side, it had a gloomy and unwelcoming look, but the part that joined Prince of Wales Road widened into a cheerful set of well-kept Edwardian terrace houses with beautifully tended front gardens. Those houses, the clean beautiful ones, seemed to belong to a different neighbourhood; in fact, a different world.

Adah's new landlord's humble abode was situated in the middle of the gloomy part which was shut off from the cheerful side by a large building curving right in the middle of the street 'as if it were determined to drive the poor from the rich; the houses from the ghetto, the whites from the blacks. The jutting end of this building was just like a social divide; solid, visible and unmoveable.' But the house, she says, needed no description since all she needed to do was to enter the street and ask for the black man's house, which looked the oldest in the street, sandwiched between two large houses owned by some Greeks. As we will see shortly, Adah's descriptions of an inner London borough are not of a London that is remembered and yearned for now. Even then socio-spatial divisions produced differing senses and experiences of place.

Princess Street (name disguised for anonymity) is a shopping street with a market running down the middle, in an inner-city area of London which has been designated as an area for regeneration under the Neighbourhood Pathfinder scheme established in the Neighbourhood Renewal Unit of the Deputy Prime Minister's Office to assist the poorest localities in the UK. Though formerly a predominantly working-class area, it was not until recently marked out as disadvantaged. The total population of the locality in 2001 was 17,200. It is an area of contrasts with 1960s and 1970s blocks of high-density modernist council housing interspersed with small roads of expensive owner-occupied terrace houses. The two are so close that it is geographically impossible to separate them. Census data reveal that the ethnic composition in the locality shifted from being approximately 80 per cent white in 1991 to 73.2 per cent in 2001. About 18 per cent of primary and 11 per cent of secondary school pupils were refugees or asylum seekers (Neighbourhood Delivery Plan: 6). Socio-economic factors reinforce a picture of disadvantage. Nearly one in ten households is headed by a lone parent, over 5 per cent of the population are unemployed (national average 3.4 per cent), nearly half (43 per cent) are in council housing (13.2 per cent nationally). Nearly 30 per cent of residents claim Housing Benefit, 17 per cent Council Tax Benefit, and 14.2 per cent Income Support compared to 14 per cent, 12 per cent and 10.6 per cent respectively in the borough as a whole. Health statistics are another representation of this inequality with a substantially higher rate of teenage pregnancies, and a higher incidence of respiratory diseases, injuries and mental health. Contrary to residents' perception, as we see shortly, crime rates are not significantly higher than elsewhere, though the rate of racially aggravated crime in the area as a whole is higher than the national average (21 per 1,000 residents as compared to 0.4 per 1,000 nationally). These figures clearly reflect a pattern of multiple disadvantages concentrated in the locality.

Princess Street runs across the middle of the regeneration area. At one end it consists of three-storey, mainly gentrified housing. Its middle part, the site of this account, comprises 42 shops including four small supermarkets, two chemists, three newsagents, a DIY shop, one butcher, one baker, a video shop, a betting shop, two dry cleaners, an insurance broker, two pubs, two cafés and a public library. There are also some empty premises (16 per cent) which, combined with some rather sad-looking businesses and wire grids on many of the shops at night, make the street look rather run down. On two days of the week there is a market which extends almost the length of the shopping street. The occupancy level for the stalls is 69 per cent on the weekday market and 79 per cent on Saturday; the number of stalls has diminished considerably during recent years. Over three-quarters of local households visit the market, of whom a quarter consider it poor. Spend per visit is low with 56 per cent spending less than £10 and a total of 88 per cent spending less than £20. Comparative questioning of 'wealthier residents from the area to the west' found that only 33 per cent ever visited Princess Street and 57 per cent were not familiar with the shopping area. The other end of the market consists of a refuge for the homeless and a children's nursery. While the street market is regarded as a reason for visiting the area, 57 per cent of the retailers thought that turnover had worsened in recent years.

Princess Street, then, is a locality where a high concentration of residents are experiencing multiple levels of deprivation and poverty. This is a street market which, unlike many such markets in London used by local gentrifiers exoticising and celebrating difference (May 1996), is largely overlooked by the middle-class gentrifiers of the surrounding streets. The area is perhaps simply too shabby to attract them. The shop owners and stall holders, though themselves better off than most of the people they serve, articulate most publicly the general malaise and discontent which pervade the community at large. It is thus their narratives that are articulated here.

For the locals who remain living, owning businesses or working in the market, there is a prevalent sense of disappointment and frustration at what is perceived as a decline in the locality's prosperity, vibrancy, moral certainty, social harmony and sense of community. A local estate agent's description of the street as 'home to one of . . . London's largest street markets . . . [where] the atmosphere has an amazing buzz' would not be recognised by those who live and work there. This decline of the market is variously laid at the door of large capital moving into the retail sector – notably the arrival of Safeways and Sainsbury's in the locality – the council and the local police. In this sense, then, it is clear that the resentments expressed by these shopkeepers towards the new migrants and asylum seekers are strongly related to their position of economic and social vulnerability in a globalising world (Betz 1994; Wells and Watson 2005). Thus local and national government are seen to have supported large capital at the expense of the 'small people'. At the same time a very real reduction since the mid-1980s in welfare benefits and resources gained by the white working class in inner-city areas governed by old Labour, combined with a view, propagated by the tabloids and media generally, that asylum seekers are gaining access to a whole slew of benefits and policies denied to longer-term residents, provides fertile ground for resentment towards new arrivals to the area.

These are very real material concerns. What I am interested in exploring centrally here, however, is the place of affect, imagination and nostalgia in fuelling people's arguably legitimate sense of being left high and dry by the forces of change. For in this hollowed-out space, abandoned by many of those who could move to the reputedly more salubrious outer suburbs of London, has emerged a nostalgia for the halcyon days when people came from far and wide to the market, when there was laughter and life in the street, and when there was a strong sense of community. While this nostalgia reflects real losses and new forms of poverty and exclusion, what these nostalgic discourses mask are the social divisions (particularly racialised divisions) of the time. Or to put this another way, nostalgia can perform the function of whitewashing the past and of producing a collective memory that reflects only some people's lives. In this romanticised vision of the past in Princess Street, the new population of asylum seekers and new arrivals (from Turkey, Afghanistan, Kenya, Vietnam, Iran, China and Bangladesh) become an easy trope for the dissatisfactions of the present. In such a context, the possibilities of enchanted urban encounters and co-mingling across differences in this public space are greatly diminished.

Striking in the research was that the majority of the white people interviewed in the market did not now express hostility towards the racial/ethnic transformation of the area that took place in the 1960s–1990s and many recounted relatively positive stories about how different groups of people had historically got along. In other words, many people romanticised an earlier time of harmony, while also being unaware of how their own comments were steeped in a naturalised sense of superiority even as they spoke of there being no racial tensions in the market.

This interview with Larry (a white shopkeeper) is fairly typical:

> The racial problem round here, that's OK. I give it very high marks for the blacks, the whites, the Pakistanis, the Muslims. I don't think I've seen no racial tension round here at all, and they seem to mix all right. Whatever it is we seem to mix all right. I treat everybody the same. . . . The racial situation is very good. I hope it stays like that. Occasionally you get a few of my customers saying 'oh you've got all these Pakis coming in' an all that. It's only natural. You get these English people or British people . . . people that were born here. . . . Since I've come here we've got two mosques . . . and I've noticed it's more black faces, more robes, whatever you wanna . . . and a lot of them . . . a lot of them can't speak very good English, but I've got a lot of patience . . . I was brought up with Indian kids. I know how to deal with them and how to speak to them. I'm allright. I haven't noticed any racial prejudice.

Asked about the changing ethnic mix, Ian, one of the white stall holders, replied:

> We look after our customers. I been serving people 30 years, I know what they want. . . . It's worked out well – different varieties of food, we serve pumpkin now, garlic, different salad-stuff – changed what people eat – 30 years ago we had six or seven lines – carrots onions apples . . .

Asked whether people get on well, he said:

> Yeah, no problem at all – 50 per cent of our customers are Asian Caribbean – I've got to know what they want . . . no problem at all. On the whole no problem at all. Mr Khan (a Sikh) – he used to have a shop here, used to sell all the West Indian fruit, where the carpet shop is – he's now got a restaurant down in Covent Garden. . . . Used to sell all the West Indian food – he comes up twice a week to get his fruit and veg for his restaurant.

Plate 3.2 Customers at a shoe stall, Princess Street Market.

Such comments were frequently articulated in the street by the traders and shopkeepers. But what is notable in this stall holder's comments, as we shall see shortly, is that these present-day narratives write out the sense of difficulty and discrimination that people who are not white British have experienced (Wells and Watson 2005). As Adah's story reflects, black people's experience in these remembered days was embedded in complex layers of discrimination, not only attributable – in her view – to the resident white community:

> Thinking back about her first year in Britain, Adah could not help wondering whether the real discrimination, if one could call it that, that she experienced was not more the work of her fellow-countrymen than of the whites. Maybe if the blacks could learn to live harmoniously with one another, maybe if a West Indian landlord could learn not to look down on the African, and the African learn to boast less of his country's natural wealth, there would be fewer inferiority feelings among the blacks.
>
> (Emecheta 1987: 76)

This picture painted by Emecheta suggests that racial divisions were as much part of the social fabric of inner-city London in the 1960s as they are now despite people's memories of the contrary. What has changed is the group to whom racialised discourses are attached. As noted earlier, the school statistics in the Princess Street locality indicate that an increasing proportion of the population are recent migrants. It is clear from the interviews that it is asylum seekers who are now seen as a legitimate category where discourses of racial exclusion and 'otherness' can be mobilised. Ugandan Asians coming to the UK in the 1970s were similarly used as a scapegoat for numerous social and economic ills. In these racist articulations some of the white people interviewed also included longer-term black residents into the discourse of 'us' as a rhetorical device to protect speakers from the racist implications of their own discourse.

Ian again:

> there have been a lot of problems since the asylum seekers come here . . . people are resentful, they get everything given to them. . . . Old age pensioner needs something . . . hasn't got a chance in hell of getting nothing. . . . West Indians feel the same, the old ones, why should they come and get new carpets, new television everything – they got £1,000 from the government starting off – last two or three years. . . .

Asked if that is the basis of resentment he said 'Yeah, that is it, yeah.' Yet of course these claims are absurd.

Such racist discourses were central to many of the nostalgic narratives. The mobilisation of black British support for these views was frequently deployed to legitimate them and exonerate the speakers of guilt. Similar views were indeed articulated by some (but not all) of the local Asian British, who, it could be argued, strategically identified with the powerful exclusionary discourses of the local white British, separating themselves discursively and practically from the powerless space left for newer arrivals. As the new owner of a children's toys and odds and ends shop (which had been run since the Second World War by a white working-class couple), Sam, a British Kenyan, said:

> Too many refugees. . . . The people that are legally here, they're made to go and bring in refugees, no jobs, they get everything, all the facilities, they don't even speak English. Too many foreigners. All my family have been British for generations.

Reflected here too is the possibility of decoupling Englishness from a seemingly compulsory whiteness (Back *et al.* 2001: 254). These comments are mild compared to others, as we see later. During two quite lengthy interviews with the butcher, which were littered with explosive racist comment, he cheerfully served and joked with the black people who came into the shop, reflecting the co-existence of 'racist talk' (exclusion) with 'colour-blind' inclusion, no doubt partly produced by the economic imperative to keep their custom. But what was consistently reiterated was a sense that the new wave of migrants to the country – homogenised discursively as asylum seekers – were all gaining access to benefits and special treatment denied to the longer-term residents of the area, and rumours of mythical gains were clearly circulated and exaggerated up and down the market (Wells and Watson 2005).

NOSTALGIA

In his recent book, Paul Gilroy coins the notion of 'post-imperial melancholia' – a 'guilt ridden loathing and depression that have come to characterise Britain's xenophobic responses to the strangers who have intruded on them' (2004: 98). In Princess Street, such a bleak account would only hold for the smallest minority of those I interviewed. More common though were nostalgic references to the market and the community in its earlier days. Clearly the market was a more vibrant place twenty or thirty years ago, and by all accounts a site of sociality and connection between many of the people living there. However, this nostalgia, I suggest, masks and writes over the social divisions of the past, and leads to a negative and disappointed sense of the present imbued with a blame culture, a celebration of the past and a fatalism about the future, where big business, the council, the police and asylum seekers are all seen to have contributed to the decline of the area. Such nostalgia appears also to act as an inhibitor to change, regeneration and new forms of sociality.

Nostalgia, a word originally coined by the Swiss physician Johannes Hofer to describe the symptoms of homesickness experienced by Swiss mercenaries fighting far from home, comes from the Greek *nostos*, to return home, and *algia* meaning a painful condition – hence the idea of a painful yearning to return home (Davis 1979: 1). Intrinsic to nostalgia is the ability to filter out unpleasant memories or, as Meerloo put it, 'nostalgia is memory with the pain removed' (quoted in ibid: 37). This pain which inclines one homeward, as it were (Steinwand 1997), is connected to some remembered feeling of wholeness and belonging or, to put it in Steinwand's words, 'the homeward pain of nostalgia presupposes that one's present place is somehow not homey enough'.

There have been different sociological approaches to nostalgia. Implicitly, writers like Weber, Tonnies, Durkheim and others shared the view that earlier societies – particularly those characterised by rural, or non-urban, sometimes suburban, life – were different, this difference being represented as a loss. These earlier arguments resurfaced in the latter part of the twentieth century in conservative nostalgic formulations of community which characterised the imagined loss of community as the source of all social ills. Thus, nostalgia operates with an idealised imaginary of the past. Yet as Raymond Williams (1975) forcefully reminded us, such nostalgic formulations served only to obscure the inequities and miseries generated by the rise of industrial capitalism and the consequent urbanisation. Typically in these formulations public life is characterised as rational, impersonal and abstract, with the family or private sphere as the place of connection, support and interdependence. In response, radical/critical sociologists and others have tended to critique nostalgic discourse as a vehicle for maintaining the status quo, for resisting socio-cultural change, and for denying the inequalities and social divisions of the past. Princess Street Market illustrates this well.

For those who can claim a stake in the past, nostalgia plays an active role in the construction of nation and a sense of belonging, leaving newer arrivals excluded from a narrative of continuity in place. As Raphael Samuel (1989: x) pointed out, the division between the established and the outsider has been a consistent feature of national life whichever period one looks at. This too becomes a route through which racist discourses are deployed and was certainly evident in the market. Chase and Shaw (1989: 2–4) specify the conditions under which nostalgia may emerge: the modern view of time as linear with the future undetermined, a sense that the present is deficient and evidence of this available from the past. Nostalgia is experienced, in their view, when there is no 'public sense of redeemability through a belief in progress' (ibid: 15). David Lowenthal similarly sees nostalgia as rooted in present dissatisfactions: 'a perpetual staple of nostalgic yearning is the search for a simple and stable past as a refuge from the turbulent and chaotic present' (1989: 21). Certainly in Princess Street Market these conditions underpinned the sense of place and the disappointment expressed at the changes that had occurred. These yearnings can be framed thematically around crime and safety, an imagined earlier cosmopolitanism, and an emptying out of the area.

Crime and safety

Despite the crime statistics and police reports indicating otherwise, there was a prevalent sense that the neighbourhood had become more dangerous. An Office of Public Management (OPM) survey (2002), for example, found 46 per cent of the local population saw drugs as a major problem, and only 42 per cent of residents felt safe out alone at night compared to 67 per cent nationally. The owner of one of the shops, Dick, expressed it thus:

> Forty years ago it was rough here but they don't shit on their own doorstep. It was honour among thieves. What ever you did, you did outside the manor.

> Whereas now they're doing their own, they're killing their own, robbing their own. It's all mixed up. You could go to old Sid down the road and he'd say, 'Leave it to me, I'll sort it out.' You can't do that now. If you had any trouble and you lived on the manor you could go to someone and they'd sort it out. You can't do that any more, because it's all mixed up with idiots, druggies, young kids who don't know what they're doing. And there's no morals. They've gotta get whatever they want and they don't care how they do it, they don't care who they mug, for whatever reason. I knew about four or five families here, and I could turn to any of them families, they'd have a word, and they'd leave them alone. I know the rules. A gospel rule you learn around Princess Street, it don't seem to apply now because they're doing each other, you do not grass and you do not grass to the old bill.

Even criminals are nostalgically recalled as being nicer then.

Sam – the British Kenyan shop owner – put it this way: 'Twenty years ago, England was a "fantastic country". Now crime, unemployed, "too many refugees".'

Even a more recent arrival to the market (though not the UK), the owner of a South East Asian supermarket, agreed:

> In the past more people used this area. These days they don't feel safe to come. It's just general safety [i.e. not racialised crime]. . . . It's not like before, the white people took care of the white people. There's a gap in the relationship. Between the whites and Asians.

And Larry, describing the time when he owned one of the market's five butcher's shops:

> One day I was opening the shutters on the butcher's shop. A little boy about 12 kept pulling the shutter down. I said, 'Don't keep doing that you're going to break it'. I went back in and I looked, he's doing it again. I said, 'You fuckin cunt you're going to break the thing'. He's standing there, he says, 'Go on, fuckin hit me'. I thought, shall I? One bash. And he's blackmailing me, this little kid. One of the police says to me, 'You all right?' . . . I says 'Can't you nick him?' . . . And I thought that's it, whereas that boy should have been scared of me, I was scared of him.

For many people, the answer to this perceived increase in crime, violence and lack of safety was the placing of CCTV cameras throughout the street, and considerable hostility was expressed at the ease with which the richer parts of the borough had managed to have cameras installed. This desire for increased surveillance and voluntary submission to the panoptic gaze of the camera instilling self-regulation and imagined protection follows a trend in cities worldwide. By giving over the problem to an unseeing but ever present eye, local citizens lose any sense of their own empowerment that alternative forms of local crime control might allow. This is illustrative of the general malaise and sense of powerlessness the people working and living in Princess Street articulate.

COSMOPOLITANISM AND SOCIO-ECONOMIC DECLINE

Intriguingly, for some of the people interviewed there was an imagined time of greater cosmopolitanism – before ethnic groups lived separately from one another. The white Jewish optician described it thus:

> Ten years ago the market was buzzing. It was very cosmopolitan, it's not as cosmopolitan as it was. As people can, they move out, people come in at the bottom end . . . it's not conflict between people, it's . . . here you tend to have pockets of Somalis, even the Anglo-Saxons. They don't interact. There's not any conflict, there's no interaction. You get people coming in [the shop] who don't even realise they're neighbours. People have their Somali evenings, Bangladeshi evenings . . . I don't know if they mix. . . . When I was a kid my family lived in the East End of London, you had British, Jewish, whatever. They all had their own little areas, but they also intermingled, they knew each other and they got on. Wherever you came from you came there to get away from some form of oppression, be it economic or political. So everyone had the same goal, they wanted to get on in life they wanted to achieve something.

Or as Ian said: 'There used to be lots of Jewish traders here – different atmosphere here – a bit of a laugh – they was more in pots and pans, toys everything . . . right the way through the week.'

Asked, how people got on, he said 'No punch ups was there? No druggies, no junkies, muggings. . . . Rougher, the whole area is, everywhere you go . . . drugs are the main problem . . . '

It was apparent that these memories of a happy intermingling referred to different groups of white people, Irish, Scottish, Welsh, English and Jewish, rather than a more diverse cultural mix.

There is now a considerable body of work within feminist theory on nostalgia. Here nostalgia is linked with the feminine through the trope of a lost mythic plenitude and what is argued to be a necessary connection between modernity, alienation and masculinity (Felski 1995: 38). In this line of argument women are seen to have an affinity with nature and some notion of organic community which is now lost. More psychoanalytically this sense of longing or yearning is for a pre-Oedipal condition of originary harmony which is associated with the maternal body (ibid: 39). Gallop sees nostalgia as intrinsically bound up in woman's desire and sense of regret, as she looks back to what she never had and encounters not a sense of loss but a sense of regret. This memory, Gallop argues, is constitutive of nostalgia (Gallop quoted in Jacobus 1995: 19).

These ideas may sound rather removed from the site of Princess Street Market. But such ideas of plenitude, or organic harmony and community, were all to be found in the ways in which stall holders and shopkeepers talked of the past. This is not to say that these memories did not reflect actual ways in which the market was different; rather, that the remembered abundance of fruit and vegetable stalls in earlier times fuelled feelings of loss. Buchi Emecheta's description of the market in her story from the early 1960s resonates with the stall holders' and shopkeepers' memories. As Adah sets off for the market one morning:

> People were passing her this way and that, all in colourful sleeveless summer dresses, one or two old dears sitting on the benches by the side of the Crescent in front of the pub smiling, showing their stiff dentures. . . . She walked into the Crescent where the smell of ripe tomatoes mingled with the odour from the butcher's.
>
> (Emecheta 1987: 182)

The general look of Princess Street, like many such markets in London now, is fairly decrepit with the old-established shops interspersed with budget shops and small supermarkets of various kinds, while in everyone's account this had been a lively vibrant trading place as long as they could remember. A similar nostalgia was expressed for the days when people went in and out of each other's houses, or controlled the bad behaviour of children simply through knowing who they were and where they came from. This, then, was a community which did not need external regulation or intervention for people to belong and feel safe, reminiscent of Wilmott and Young's (1957) account of Bethnal Green.

There was instead a discourse of the area having emptied out with the decline of the market as a trading space. When Ian started to work in the market as a boy of 11 in about 1960 there were 24 fruit and vegetable stalls and four butchers; now there are two and one of each respectively. The South East Asian shopkeeper gave this explanation:

> [it's] the economic situation. The people living in this area, Asian people from . . . Bangladesh, all these people bring a kind of mixed cultural thing. Mixed neighbours. It's not like before, the white people took care of the white people. . . . There's no industry, no banking services. This kind of thing makes the area look forgotten in some way. In the past more people used this area. These days they don't feel safe to come.

Sainsbury's opened its second grocery shop in Princess Street in the early twentieth century, and though it now houses another business, people point proudly to the old green JS tiles which can be seen through the window.

Respondents commented on how the area had felt like a stable community steeped in tradition where children grew up in the market and shopped there themselves as adults. Larry, who owns the freehold on his shop and the shop opposite, described it thus:

> You've got a market that's been here for hundreds of years. You've got a generation that's been coming here for hundreds of years. The younger ones

> are getting used to going to Sainsburys. The amount of people is getting less and less and less. I had the butcher's across the road, I bought it for the flat upstairs. I said to the council why you opening that big store . . . that's in walking distance of the market, and you're going to have a bus. That's wrong. You've got five big stores in [the area]. . . . It'll be a nail in the coffin of the market. The Council gave planning permission to Safeways cos they were going to bankroll the flats. . . . Because Mum used to come here their kids come here. Go down to the market you'll get what you want. Safeways opens. That habit's broken. I sold [my shop]. A quality butcher's, he'd been serving his customers for hundreds of years. We had people coming in from [all the wealthier areas nearby] because it was good-quality meat. Once that went we started losing the people. If they want cheap meat why they going to come into the market. They're car people. Mr Meat, he'd been here maybe 100 years, he closed down. There's not enough people coming into the market.
>
> Councillors come round here at election time and they say they'll do this an they'll do that and they do bugger all. . . . The market needs advertising to bring people in. More stall holders. More noise and hassle, that's a market. That's its character.

During the interview one of his customers (who owns the cycle shop) chipped in:

> The reason the market's declined is Safeways. Why should they come shopping down here . . . it's cheaper [there]. Markets are finished. And the stall holders that are in the market now are not real stall holders. Stall holders years ago, it was their job. I've got a picture indoors of the market in 1880s when it was a proper market. It was a business. The stall holders that are here on Thursdays and Saturdays are people on the dole. There's about half a dozen stalls that 'ave got all erm, eh, careful what I say, illegal refugees selling stuff. There's two or three proper ones. The fruit and veg stalls. They've closed half the market anyway. When I lived here we had three Sainsbury's, a Woolworth's, Boots the Chemist. That shop there is Sainsbury's second shop. [Our business] could not survive on local business. I've been there 52 years.

As a result, many people reported a desire to leave. As the Sunny Café waitress put it: 'If I could move I'd get out of here. . . . Even the kids are polite to you over there. The names you get called by these kids is unbelievable'.

Felski's (1995: 41) discussion of the prevalence of mourning for an idealised past during the nineteenth century as a destructive desire to regress clearly has some bearing on Princess Street today.

This picture was borne out by the regeneration co-ordinator. From his point of view, the market had been a major area of concern for some years with considerable sums of money having been spent on its development in previous regeneration initiatives to little effect. He argued that the major problem was poverty, with the result that there was a poor range of shops and goods in the market and limited money spent in the locality, while people with a disposable income tended to shop elsewhere:

> The reality is quite lot of the people in [the area] who want it to be revived have some sort of vision of it being like markets of old – sort of happy white working-class people going to doing their shopping without recognising that people have changed and their economic position has changed and the racial make up has changed, and the retail shopping patterns have changed enormously and there ain't no turning that clock back. And my feeling is that the council aren't really committed to trying to change that. . . . A lot of ideas have been suggested over the years – flower market, French market – that has never been pursued by the council. My own take is that the market is failing cos you have poor people shopping there – you have an agglomeration of low-quality low-cost shops there that put people like you and me off shopping there.

Embedded in these accounts is an imagined time of easy encounters with others, even those of different races and ethnicities, which has now disappeared. What I am suggesting is that this nostalgia operates to close down the possibility of accepting the social and cultural changes to the locality, thus mobilising a focus on the problems of difference, rather than finding ways to negotiate the new cultural practices engaged in by different racial/ethnic groups. Resistance to difference was focused on two issues: the lack of shared language and the lack of shared religion. These two elements appeared central to the established residents' notion of a British identity – hence the ability to fold the earlier black migrants to the locality into the category of Britishness. In this neighbourhood the collapsing of 'white' skin into English identity – which Ahmed (2000) stressed as crucial to insider status – has thus apparently started to unravel, although the inclusion of some black people seemed to be used to reinforce the exclusion of other black and Asian people (e.g. Somali Muslims and Bangladeshis).

Being Anglophone

A persistent theme of the shopkeepers' 'talk' was the problems presented for community formation by the absence of a shared (English) language between participants in public space. The optician – who hoped that he was not racist, 'I can't be' – argued that:

> you should be able to speak the language, and why should the road signs be in any language other than ours [referring to the Indian language street signs in Bethnal Green]? . . . I resent it. Why should it be that way? . . . Our language here is English, therefore everything should be in English.

This is not, it should be noted, simply an issue of communication. Space, as expressed through street signs, should be transparent to him, as an English

Plate 3.3 Princess Street Market, Saturday morning.

reader, in a way that it would not be to non-English or bilingual readers. It is not simply a question of a common public language, but a denial of the legitimacy of people speaking other languages to shape English public space. His resentment of the accommodation of non-English speakers is echoed in the butcher's lament that 'They won't speak English', and in one of his white customer's complaints that her grandson had come home from school with Bengali writing on his hands.

That's 'Not right,' she exclaimed in a tone that evidently was meant to convey the unassailable logic and commonsense basis of her condemnation. This dismay at the loss of English as a shared public language is not confined to white people; the British Vietnamese owner of a small supermarket commented that:

> Asian people from Bangladesh, all these people bring a kind of mixed cultural thing. . . . Some of them can't even speak English. You can't understand from them, they can't understand you, it makes the situation even worse . . . [the] buying–selling relationship is worsened by language.

As was noted earlier, Sam, the British Kenyan shop owner, stressed: 'they [refugees] get everything, all the facilities, don't even speak English.'

BEING CHRISTIAN

Protestantism, Linda Colley (2003) has argued, was crucial to the formation of British national identity. While Christianity as a religious practice may no longer be so important to defining what it means to be British, as a cultural marker it continues to frame the borders of inclusion and exclusion (Parekh 2000: 242). In particular a sharp rejection of the right of Muslims to claim British identity is apparent in many of these interviews.

Namaad and Pam own an ex-catalogue shop on Poppy Street. Pam works as a waitress in the next-door café (a 'greasy spoon' owned by Peter, a Cypriot). Pam commented that 'People don't like Muslims'. A shop on the next turning has been converted into a mosque, and this had caused 'a lot of trouble':

> They didn't like the idea when all these Muslims opened the shop and made it into a mosque around the corner. It wasn't that it was Muslims. It was just that 'cause it was so small, they used to sit on the pavements with no shoes on praying and you've got to cross over to the other side of the road. Lots of complaints about that. People would moan about that. Not 'cause they were praying, but 'cause there were big gangs of them sitting on the pavement, no shoes on, you got to cross over and cross back to the shop.

In response to a question about whether or not people are hostile towards Muslims, Pam commented that, 'I think they are. It's got worse since all this American thing. You get nasty remarks. Especially about [inaudible]. I say: "Well, it's just their religion the way they do their food. It's their religion innit." "It stinks. It's dirty [they say]".'

This rejection of the right of non-Christians to shape public space is not confined to resentments against Muslims. A customer in the butcher's, a white working-class woman in her sixties, in the midst of a racist tirade about the imagined entitlements of 'them', complained that, 'My grandson's celebrating not Christmas but Hindus [at school]. Teach them their own religion first.'

Plate 3.4 Old shop converted into a local mosque.

This construction of non-Christian religious presences and practices as negatively reconstituting public and institutional space was also implicit in several respondents' otherwise rather puzzling preoccupation with butcher's shops. The area has a history of meat retailing from both shops and stalls, the first having been established in 1859. The butcher's comment – that there 'used to be five butcher's shops here, now I'm the only one . . . they've got halal this and halal that' – explicitly ties the decline in meat retailing to the increase in a Muslim presence. John does not make this connection explicit, but his comments also connect the loss of meat retailing to the decline of the area and its inability, in his view, to attract 'good' incomers. The comments of one of the self-proclaimed racist couple who own a small hardware stall indicate the sense that butcher's shops in this neighbourhood signify a broadly homogeneous linguistic, cultural and religious community. In choosing rabbit as the signifier of this lost age, he does not just refer to a type of meat but to one strongly associated, because of its cheapness, with working-class cooking, and particularly with working-class English cooking, and perhaps with Christian food – rabbit is *treyf* for observant Jews and, according to some Islamic sources, *haram* to Muslims. The loss of demand for particular types of meat, and therefore of butcher's shops, is implicitly, and on occasion explicitly, blamed on newcomers who are therefore responsible for undermining the fabric of community life.

Meat preparation has long formed part of a thinly veiled racist discourse that constructs Muslims and Jews as the antithesis of Englishness. In June 2003 the British government moved to ban the halal and kosher slaughter of animals, arguing that these methods are 'inhumane'. The significance of dietary taboos in constructing notions of pollution and purity that form the boundaries of cultural groups has been a cornerstone of cultural theory since Mary Douglas' seminal work (1991). Pam's remark quoted above, that people say Muslim food (halal meat?) 'stinks. It's dirty', underscores the pertinence of these themes to discourses of exclusion and inclusion in this particular neighbourhood.

Welfare chauvinism

A discourse about who can legitimately claim to be British is drawn on to challenge the entitlement of asylum seekers to economic and political resources at the same time as challenging the right of any British nationals who are neither Christian nor Anglophone (and, possibly, not white) to claim national benefits, invoking what Betz calls 'welfare chauvinism' (1994: 173–174). Objections to 'asylum seekers' receiving welfare benefits are raised in the same discursive moment that resentments about British citizens who are not Anglophone or Christian are expressed. A customer in the butcher's shop, for example, moved quickly from her objections to her grandson celebrating Diwali, mentioned earlier, to assertions about what 'they' get from the council: 'I'm too scared to go out. They mug you. I've got a car but I won't leave the windows down. They get council flats and HOUSES [shouting]. They're getting free driving lessons and the deposit on a car from lottery.'

This collapsing together of the identities of British ethnic minorities and asylum seekers reveals the plasticity of the concept of asylum seekers, which, in fact, is used interchangeably with refugees, Muslims and often simply 'they'. Thus asylum seekers are not, in these respondents' 'talk', simply 'people escaping oppression and seeking safety in Britain'. The term signifies a much wider population who, in the speakers' view, have tenuous claims to being British. The self-avowedly racist shopkeeper of a small hardware store displays this most clearly when she runs together her complaints about the presence of Muslims in public space and her resentments about what she claims 'the ethnics' and asylum seekers get. She explicitly frames her attack on these groups within a highly exclusionary notion of what it means to be British:

> They closed down L. Hall [a community centre]. Going to turn it into a mosque. They've already turned that shop into a mosque. They're lying all over the street. We have to walk in the road. I get nothing. You and me, it's our country. I don't care if you're racialist or not. It's OUR [shouts] country. They come here. Young people get pregnant, they get a flat. And the ethnics. They get a flat, carpets. The government's going to give asylum seekers £400 to buy a car and money for lessons. And £2 a day for dog food. If we've got no money we can't eat.

The butcher's 'talk' also clearly displays this move from one category of exclusion – Asian, Muslim – to another – refugees and asylum seekers:

> They get jobs, houses. If you're Muslim you go to the top of the housing list. They took over Springfield and now they're taking over here. They won't speak English. Why don't they want to know us? Bob [Asian shopkeeper] wouldn't [hypothetically] let his daughter go out with my son, but I'm the racialist. If you're English, white English or black English, you get left out. People get off the boat and get thousands straight away.

'They' in this extract refers to the relatively longstanding Bangladeshi community in Springfield, many of whom settled in Britain as adults in the early 1960s under the voucher system of immigration, being joined by their children in the early 1970s. Despite the longevity of their residence, they remain, in his view, not English. In his reference to the boat Fred links the most recent arrivals to the UK, asylum seekers in 2003 – who mostly arrive by air or land, with the Caribbean settlers of 1948. He quickly acts to correct this unveiling of his racist discourse by immediately adding, 'If you talk to old West Indians [note the colonial West Indian rather than the post-colonial Caribbean] they would agree – they know what it's like now'. References to asylum seekers or refugees are used to (barely) conceal the racism that underlies the speaker's resentment.

As the invocation of this mythical 'West Indian' figure suggests, many of these respondents are at pains to deny the racist provenance of their complaints. The butcher, for example, begins his racist tirade with the words, 'How can I say this without sounding racialist?' and later, in an astounding claim, he says, 'If you call your neighbour a black bastard or a Jew bastard that don't mean you're racialist. The black neighbour don't think you're racist. It's the do-gooders.' Interviewer: 'Who are they?' 'They work in the Town Hall, in the government. They're not civil servants. They're do-gooders.' This strategy of denying that they are racist, and expressing resentment at being thought of as racist, while expressing implicit or explicit racism was adopted by a number of the white respondents. They seemed to recognise that being racist undermined the credibility of their grievances. Albert, an old white man, who sat in John's shop throughout the interview said, on being asked by John what he thought about 'the racial problem':

> Next door I've got a coloured woman and two boys. I know they're young. Last night about three o'clock in the morning music blaring out through the open windows. But you can't say nothing or they'll call you a racialist. You've just gotta stomach it. What you do? If you complain 'oh what you're racialist', you just accept that they're here and you can't do nothing about it.

In each of these extracts from our interviews, in addition to the conflation of asylum seekers and non-Christian/Anglophone/white British nationals, the exaggeration of the benefits asylum seekers are entitled to is very striking. These discourses on asylum seekers in Princess Street are thus framed within a politics of resentment about what is seen as undue political and financial support being given to people who either are not British nationals or whose British identity is seen as compromised by their refusal to abandon the signifiers of other national identities.

CONCLUSION

This story of the market brings to mind the notion of the foreigner in Kristeva's account (1991: 103) (see Chapter 1 above) who

> signifies the difficulty of living as an *other* and with others; politically he underscores the limits of nation-states and of the national political conscience that characterises them and that we all have deeply interiorised to the point of considering it normal that there are foreigners, that is, people who do not have the same rights as we do.

More specifically, Princess Street throws light on the difficulties of encounter between longer-established residents and newer arrivals who are ethnically and racially different in old working-class neighbourhoods in the city. What I am arguing here is that in places undergoing rapid socio-cultural and economic change through the processes of globalisation, it is easy to yearn for some earlier moment which is imagined as more harmonious, plentiful, safe, settled, social and fair. The nostalgia which emerges writes out former difficulties and tensions, while in the present it is constitutive of resistance to different 'foreign' cultural practices, new norms and change. In these unsettled spaces, blame and resentment can thus erupt towards those who are different, and who are often perceived as beneficiaries of privileges denied to oneself. In other words, a sense of loss is translated into envy and resentment towards those who are reputed to be its cause. Though there was plenty of evidence of sociality and connection in the market, my point is that rubbing along across differences in public space, in a context like this one, is thwarted and inevitably constrained. If since the mid-1980s the local council and other key players in the locality had played a more active and strategic role in tackling the socio-economic decline of the market, the story of this market as a space of more enchanted urban encounters, could more straightforwardly be told.

NOTE

The statistical material is drawn from several local government sources which are withheld here to preserve anonymity.

4 Risky space and money talks

The Hampstead ponds meet state regulation

The new risk culture of the late twentieth and early twenty-first centuries has produced a heightened awareness of the imagined and real dangers of moving freely through city spaces. Fears of bodily harm from pollution, traffic, violent others, road or rail accidents act to inhibit the urban subject from enjoying public space as a space of freedom, encounter and expression. The proliferation of media accounts of the dangers lurking outside the front door have created a climate of fear where any accurate assessment of potential dangers can easily be ignored. Around any street corner lurks the paedophile, the mugger, the marauding teenager, the dangerous driver, the terrorist, or cancer from pollution created by cars, industry or other people's smoking.

This perception of increased risk at the level of the urban subject diminishes the possibilities for sociality and encounters. This is well illustrated in the everyday lives of older and younger people, as we shall see (Chapters 6 and 7 below). For Furedi (1997: 9) this evaluation of everything in terms of risk is the defining characteristic of contemporary society, where what is striking is not the level of insecurity, but rather the deeply conservative way in which this condition is experienced. In his view risk has become a profoundly significant factor organising the ways in which we conduct ourselves in everyday life:

> The disposition to perceive one's existence as being at risk has had a discernible effect on the conduct of life. It has served to modify action and interaction between people. The disposition to panic, the remarkable dread of strangers and the feebleness of relations of trust have all had important implications for everyday life. . . . Through the prism of the culture of abuse, people have been rediscovered as sad and damaged individuals in need of professional guidance. From this emerges the diminished subject; ineffective individuals and collectivities with low expectations. Increasingly we feel more comfortable with seeing people as victims of their circumstances rather than as authors of their lives. The outcome of these developments is a world view which equates the good life with self-limitation, and risk aversion.
>
> (ibid: 147)

Douglas and Wildavsky (1983) have similarly highlighted society's new attentiveness to, and subjective consciousness of, risk, which often bear only a limited relationship to an increase in real dangers.

It is arguable whether risks have actually increased in late modernity or whether there has been a shift in their constitution. Beck (1992) contends that in pre-modern society there were only hazards, not risks. For him, processes of rapid globalisation in late modernity have proliferated risks which transcend national boundaries and require new forms of calculation and intervention. Similarly for Giddens (1990, 1994, 1999) there has been a shift from the prevalence of external risks – those associated with tradition and nature – to manufactured risk, hazards that are produced by ourselves and our interventions into the conditions of nature within a context of increased globalisation, so that governments are now caught up with new forms of risk management (Giddens 1999: 33–34; 1994: 4).

It is the new modes of governance – different risk regulation regimes (Hood *et al.* 2004: 171) – which are now deployed and their effects which are of relevance to the story told here. Rose points to the emergence of new forms of governance through the freedom to make choices of one's own which not only multiply the 'points at which the citizen has to play his or her part in the games that govern them' but also 'multiply the junctures where these games are opened up to uncertainty and risk, and to contestation and redirection' (2000: 95). In his account citizenship requires urban subjects to participate and act in their environment, making choices about their lifestyle, reshaping their environment, engaging in complex circuits of power and governing themselves. Thus risk reduction becomes part of the moral responsibility of urban citizens themselves (ibid: 103). Risk in these terms represents a way of ordering reality and rendering it into a calculable form (Dean 1999: 177). Such an understanding frames the narrative told here of the attempts to curtail winter swimming in the Hampstead ponds in North London. It is also a story of how governments, local and national, are driven to protect themselves against potential responsibility for actions and events under their jurisdiction which produce harm.

Also through this story I aim to highlight another way in which state practices and strategies have impacted on the provision of public space, because in this site the two forms of intervention became closely intertwined. Here I am referring to the more familiar tale of the withdrawal of public facilities and space, legitimated by discourses of insufficient public funds available to local authorities for expenditure on seemingly less essential services. In the UK it is estimated that £73.4 billion were spent on local services in 2004–2005. In line with Labour government priorities, councils spend the largest proportion of their budgets on education (40 per cent) and social services (19 per cent) (local.gov.uk, March 2005). Initiatives to improve or sustain public space are inevitably lower down on the list of priorities in the wider context of deprivation indices. Their lack of visibility, on the one hand, and the difficulty of assessing their social impacts and benefits, on the other, mean that public space provision is rarely seen as a key area of concern. Nevertheless, between 2001 and 2006 £1.9 billion have been allocated by central government to the Neighbourhood Renewal Fund for the

88 most deprived local authorities in Britain with the objective of decreasing the disparities between poorer and richer areas. But the proportion of this expenditure spent on public space even in these areas is likely to remain small in the face of competing demands for expenditure on housing, employment and urban regeneration initiatives.

Over several decades there has been a shift towards the privatisation of many leisure services, combined with a growth in the provision of private facilities such as local gyms, swimming pools and children's play areas, and these have attracted higher-income urbanites away from the public facilities. This too has had an impact on the use of public spaces. Despite increasing accountability, accounts of continuing mismanagement of public funds, the concentration of expenditure in administrative costs rather than in the provision of services on the ground, and the proliferation of unnecessary bureaucracy also constitute an often articulated theme in public debate. In the history of the recent threats to Hampstead ponds in North London, these interwoven themes of governance of risk and an apparent (though disputed) lack of adequate funds for the provision of public space are starkly illustrated.

THE HAMPSTEAD PONDS

In Hampstead Heath, North London, three bathing ponds – the site of pleasure and eccentricity for a growing number of sun, water and 'back to nature' lovers – were little known outside London except to readers of the *Lonely Planet Guide* or tourists with an interest in quirky places. Since about 2002 these usually rather chilly spots have heated up as the state, in the form of the Corporation of London, intervened to close them down, driven by budget deficits and a fear of its potential liability in the event of accident. Barely visible except to the *cognoscenti*, these natural water bathing ponds – one for men, one for women (which are the focus of this chapter) and one mixed – are fed by an underground river. The ponds represent three public spaces, each with its particular ambience and regulars, which have been revered and enjoyed by locals and visitors since the 1880s (the men's pond) and 1924 (the women's pond). The special quality of these sites has inspired fierce support for their preservation over the years and eulogies which possibly have no parallel for any other urban green space in the city. Thus they are variously referred to as: 'a symbol of a little piece of paradise', 'a rare and marvellous natural feature in the urban landscape of London' (*Camden New Journal* (hereafter *CNJ*), 27 February 2003), 'a place of beauty . . . an Arcadian heaven' (Griswold 1998: 5), a sanctuary, where, for one writer, 'I hesitate to say it's spiritual but there's something magical and peaceful about being here' (Cane and Griswold 2002: 4), while for another 'that moment when a kingfisher flies over my head stays with me later on the sweaty Northern line' (*CNJ*, 25 July 2002, p. 21). The dominant narrative of these ponds is one of a harmonious and perfect public urban space where there is a co-mingling of strangers and friends – recalling Young's (1990) community of difference – and the pleasure of bodies free to move, breathe and play in a 'natural' environment. So popular is the women's pond that during the

Plate 4.1 The entrance to the women's pond at Hampstead Heath.

hot summer of 2003 as many as 8,000 women visited it in one day, with queues of 300 stretching back from the steps into the water. Interestingly, there have been no moves to challenge sex-segregated swimming, as has happened elsewhere. For example, in an Australian site – the McIvers Ladies Swimming Baths at Coogee beach-men fought for access, mobilising anti-discrimination legislation (Iveson 2003) to win their case. Instead, the bathers self-segregate – men who want to swim in a space without women use the men's pond and others go to the mixed pond. Occasionally men who have no interest in the ponds, but are hostile to women taking their own space, have tried to force their way into the women's pond as voyeurs.

Men having been swimming in the ponds since the 1880s. In 1883 a group of men founded the Lifebuoys, with only one rule – that members swim every day of the year in the men's pond (*CNJ*, 2 January, 2003, p. 16). The former Chair of the Kenwood Ladies' Pond Association (KLPA), Margaret Hepburn, described how her husband swam in the ladies' pond as a boy when the then landowner, the Earl of Mansfield of Kenwood, invited children to swim there. It was only designated as the ladies' pond later (Griswold 1998: 8–28). In 1924 the Kenwood Ladies' Pond, the South Meadow and the upper three Highgate Woods were acquired through public subscription by the Kenwood Preservation Council and offered to the London County Council, which gratefully accepted them as an extension to Hampstead Heath. Following the opening of Kenwood to the public by King George V in 1925 amidst a great fanfare attended by 3,000 dignitaries, including the Mayor of Paris and the Home Secretary, with a choir of 500 school children, the Kenwood Ladies' Pond was opened. In 1926 the Chairman of the LCC Parks Committee announced that the ladies' pond would be fitted with proper dressing accommodation and screening, and the pond was officially opened for women. At the time it opened it was a secret place. Kingy was the first lifeguard:

> [She] had a metal table with a tablecloth where she served cups of tea . . . would cycle from her home in Hornsey winter and summer, come hail, rain, wind or snow for twenty years or more . . . she was a strong-boned person with a ready welcome. . . . She would rub our backs on colder days.
>
> (ibid: 10)

From the women's pond's inception a sociality which mapped a norm of domesticity, cosiness and desexualised tactile relations is reported in the oral and written histories (see Chapter 5 below). Winter swimming was part of the ethos of the men's and women's ponds from the early days, with the Kenwood Regulars' Swimming Club featuring in the *Hampstead and Highgate Express* (hereafter *Ham and High*) in 1927:

> Plenty of swimming and no coddling is the advice of a score of young office girls to keep free from flu and other winter ills . . . twelve girls wearing swimming costumes of brightly coloured wool, which they had knitted themselves . . . never miss their swims in the pond all through the winter . . . on several occasions one girl has had to jump in and break the ice before the others could take the plunge.
>
> (Griswold 1998: 11)

By 1931 men formed the majority of these regular winter swimmers with women's lack of participation excused by the gentleness and weakness of their sex. Nevertheless a growing number of women were resisting what Horwood (2000: 653) describes as patriarchal prudery, inserting discourses of health in place of those encouraging modesty. A certain toughness, though, was required and the changing hut at the pond remained wooden and primitive until 1955 when the LCC built a new hut and deck. The women's pond thus was a space which produced the idea of strong non-conforming female bodies disrupting conventional notions of femininity.

Two new lifeguards took over from Kingy: Pat from 1958 to 1983, Vera from 1972 to 1987. This was an era of camaraderie, where the lifeguards did their own decorating, made tea and Bovril for the winter swimmers and celebrated the Queen's Jubilee in the cabin with wine. The ponds were becoming more and more popular, with the meadows packed solid on sunny days. According to Griswold (1998: 24) there were eight early (winter) swimmers in this period, each with her own nickname (like Amnesty Ann and Underwater Barbara) and regular time to swim. Some of these were 'Eastern European ladies' who sunbathed behind the hut (ibid: 24), and there was even a cat in residence who was fed by the swimmers.

Since the mid-1990s there has been an increasing emphasis on health and safety regulations to which the ponds have been required to adhere. This is the new regulatory climate of managing risk, where health and safety regulations are discursively deployed to construct public spaces as potentially life threatening and dangerous, thus delimiting the possibilities for embodied differences to be enacted in such sites. As Jane – the chief lifeguard at the women's pond since 1996 – put it:

> When I started there were two lifeguards. Now we have seven in the summer and in the winter – now we have to be much more safety conscious. Anything that might happen here, we have to assess the risk. All the lifeguards swim all through the winter. I have encouraged that.

What emerged was a clash between a regulatory state bureaucracy and a group of individuals who were operating with a very different notion of risk – one which falls within a regime which Hood *et al.* (2004: 13) refer to as the 'fatalist risk regulation regime'. This is a notion which stresses the unpredictability and unmanageability of hazards, and a preference for ad hoc and post hoc responses rather than a curtailment of freedom.

The conflict between the swimmers and the Corporation of London arose over the latter's attempt to curtail winter swimming which had been a tradition in both the men's and the women's ponds since their inception, with various levels of popularity (there was a notable decline in women's participation before the war (Griswold: 1998)). By March 2004, 80 or so women and an equivalent number of men were swimming through the winter. Ken Kennedy (62) of the USA described its pleasures: 'swimming there in the morning is like being in another world. Everybody is living in 24-hour lives but when you swim in those ponds it is an unbelievable release from it all' (*Ham and High*, October 3 2003, p. 4). This was echoed by the Chair of the KLPA, Margaret Hepburn (82): 'Climbing into the pond is like climbing into a painting. It's spiritual and emotional as well as physical. . . . It's very sustaining. Not to do it now is unbearable.' And another member, Mary Cane (56, writer and ex-mayor): 'It is breath-taking to stare at the pink and purple morning sky while doing the back-stroke' (*Independent*, 9 October 2003, p. 9). Breaking the ice to swim is not unusual. Winter swimming has been a long tradition also in Nordic and Russian seas.

The winter swimmers in the women's pond are a heterogeneous group though there is a clear majority of older women many of whom have, or have had, professional or artistic careers. Thus they include one (ex-)lawyer, opera singers, actresses and journalists. The men's pond has a similar profile. All the swimmers advocate persistence, swimming every day from the summer into the winter, thereby building up tolerance and resistance and reducing the risk of shock to the body. They also advocate extra clothing and moving quickly afterwards, with some people swimming in neoprene (diver's) gloves and/or socks. The claims

Plate 4.2 Two of the lifeguards at the women's pond, Hampstead Heath.

Plate 4.3 Ruthie Petrie, formerly of Virago publishers and *Spare Rib*, swimmer for 30 years.

made for it are numerous – that it helps the circulation, builds up the immune system and does wonders for health. One woman, Wynn who is 92 and arrives on her bicycle to swim twice a day in the winter, described her daily experience:

> We ladies have a bowl of hot water to stand in when we come out – it has been taken away from us from time to time on grounds of health and safety – but standing in the hot water with your hands in as well helps to get warmer quicker. In rebellion people have brought flasks and hot water bottles when the hot water was forbidden!

Hampstead Heath first acquired public space status in 1871 and became the London County Council's most popular open space by the end of the century. Fiercely defended by the Hampstead Protection Society (formed in 1897), in collaboration with the LCC, against threats of buildings and roads within its boundaries, it has remained a popular space ever since. With the demise of the London County Council Hampstead Heath fell under the jurisdiction of the Greater London Council until Thatcher's Conservative government abolished that public authority, mainly as a result of hostility to its public expenditure and participative ethos under the control of (Red) Ken Livingstone. It then passed to the London Residuary Board in 1986 and, in 1989, on to the Corporation of London (CoL) which has been responsible for its management ever since. In 2002, the CoL, fearful for the safety of swimmers in the early morning darkness, commissioned a report from the Amateur Swimmers Association (ASA). This was presented to the Hampstead Heath Consultative Committee on 25 September 2003 at the Parliament Hill Lido, the conclusion being that 'The Corporation of London had no alternative but to comply with its statutory duties and comply with the recommendations of the report'. The ASA made a number of recommendations: that swimming in the dark and breaking the ice be stopped, that there should be no swimming on foggy and misty days, and that winter swimming should be restricted to regular swimmers who would be compelled to have medicals and who would have to swim four times a week to be acclimatised by 'habituation to cold water'. Only regular winter swimmers would be allowed to enjoy the Christimas swim, which is followed by mulled wine and mince pies, to prevent the risk of hypothermia or Sudden Immersion Syndrome. If non-habituated swimmers continued to participate informally, the report concluded, 'it is inevitable that sooner or later a tragedy would occur' (ASA 2003: 19). Lifeguards were also to be compelled to take medicals twice a year and swim four times a week. The tone of the report is patronising and legalistic.

The report met with a frosty reception. After twenty minutes of trying to present his arguments about individual freedom versus CoL liability and responsibility, the Director of Legal Affairs at the Institute of Swimming Teachers and Coaches sat down, humiliated by shouts about the report's irrelevance and patronising attitude. 'We are not children,' people shouted (*Ham and High*, 3 October 2003). The meeting broke up as consultants' voices were drowned by the swimmers' expression of outrage. As the solicitor representing the United Swimmers

Association (USA, a self-selected group of the ponds's swimmers) explained: 'Instead of trying to find their way around the question of early morning swimming in the winter, the CoL seems more interested in minimising its liability.' From the Chairman of the Hampstead Heath Management Committee's viewpoint:

> The Corporation has a statutory duty of care to ensure that health and safety requirements are met for both swimmers and employees. We cannot allow swimming to take place in the dark – the safety of the swimmers and lifeguards is our priority.
>
> (*Ham and High*, 28 February 2003)

In any event the swimmers were prepared to accept personal liability, and saw the report as an intervention by what they defined as the 'nanny state', designed to curb their freedom to make their own decisions and knowingly calculate the possible risk to themselves that such swimming represented. After all, it had been done for over a century with little mishap.

In the following months the issue was hotly contested at a number of committee meetings in the local town hall and in the Corporation of London, receiving considerable media attention locally and nationally. The key points of the CoL's proposals were to limit the hours available for supervised swimming, thus curtailing the possibility of those in employment to swim on their way to work. Not everyone was sympathetic to the campaign – some locals complained of the costs incurred in maintaining the current hours (*CNJ*, 24 July 2003).

Self-regulation was proposed by the swimmers as a solution, meaning they would dispense with the early morning lifeguard and sign waivers absolving the relevant authorities of responsibility for whatever might happen. After some months of debate and campaigning the possibility of swimming without lifeguards came under discussion. After legal wrangling the proposal under consideration in March 2004 was that a small club of those committed to winter swimming might be formed for the purpose of swimming in the mixed pond across the heath from the other two ponds in the designated hours under a licence. Robert Sutherland Smith (Chair of the USA) summed up his views as follows:

> The USA was created to defend the legitimate and historic right to use the Hampstead Heath ponds for swimming throughout the year, which is a right granted by Parliament to the people. It is not subject to the manufactory economics taken up by the superintendent of Hampstead Heath. . . . If Hampstead Heath was intended to be run economically it would not exist. It would have been dug up and turned into a housing estate long ago; no doubt its uneconomic swimming ponds filled in for car parking. It was to thwart such an eventuality that Parliament passed the Hampstead Heath Act protecting it for the full use by the people. Even hard-nosed Victorians had the wisdom to know that factory economics did not justify everything in a civilised and balanced society.
>
> (*Ham and High*, 11 July 2003)

Illustrated here in Hood *et al.*'s terms (2004) are the tensions arising from different risk regulation regimes: the swimmers wanted the freedom to manage the risks for themselves, and the bureaucracy was motivated by a desire to minimise the potential dangers in advance, wary as it was of the possibilities of litigation in the event of an accident. In some sense we see urban citizens here who are willing and able to govern themselves (Rose 2000), and are resentful of moves to redirect and reshape their practices. Thus, the curtailment of winter swimming represents an extreme case of an already interventionist bureaucracy driven by concerns about health and safety, which to the users of the ponds appeared variously superfluous, patronising and unnecessary: 'People have been swimming there for years, so why stop now?', 'People swim in all sorts of places, including the sea, which are not regulated' and so on. But there is a difference here, namely that the Corporation, as the managers of this space, feared the consequences of their own responsibility. As Beck (1992) and Wildavsky (1988) have pointed out, notions and perspectives on risk are culturally and socially constructed. In Britain today there is the double fear that a litigation culture is being imported from the USA, and that an increasing shift towards the 'nanny state' – particularly under the Blair government – means that individual responsibility is being taken out of people's hands by the experts. This is a shift from the individual's 'care of the self' (Foucault 1988) to the arenas of the state. This has produced new subjectivities, vulnerable and constrained in enacting bodied pleasures in public spaces when and how they choose. This represents a shift in the exercise of power.

The proposal to negotiate a licence came to an abrupt halt in the middle of 2004. In March 2003 a vehicle accident on the Heath had led to the prosecution in the following year of the Heath superintendent who was fined £80,000. It seemed no coincidence that this coincided with the Heath managers' withdrawal from discussions and refusal to negotiate. The winter swimmers did not relent and pursued their case to the courts. On 26 April 2005 a High Court judge, Mr Justice Burton, 'struck a blow' for 'individual freedom', speaking against 'a grey and dull safety regime being imposed on everyone' with a ruling that swimmers have a right to take risks and swim in the ponds on chilly winter mornings (*The Times*, 27 April 2005, p. 25). The Corporation's refusal to allow swimming without the presence of lifeguards was, he argued, based on a misapprehension of health and safety legislation. Central to his argument was the concept of risk. Recognising that there was inevitably some degree of risk involved, nevertheless he asserted that:

> the swimmers would also be exposed to risk as they drive or walk to the pond and as they travel from the pond to their work or their homes. No one would suggest that the corporation should be responsible for an accident resulting from the risks of a traffic accident or a heart attack while walking or running to or from the pond. Risk is inherent in life and some risk is unavoidable.
>
> (*Ham and High*, 29 April 2005, p. 1)

The decision was hailed by supporters as a victory for common sense, another successful attack by ordinary citizens on the 'nanny state' and the government-sponsored cult of health and safety, and as the re-establishment of 'an important principle of personal freedom' (*The Times*, 27 April 2005, p. 25). What the decision also represented was an important precedent which had wider implications not just for all open-air swimming in England, but also in relation to the recent growth in litigation and suing, which many people saw as an unwelcome import from the USA. More broadly, it represents a legal challenge to prevailing discourses and norms of the risk culture.

A QUESTION OF MONEY?

In late 2004 the ponds came under attack from a different direction. The Corporation found that the budget was projected to be overspent by £230,000 and a similar amount the following year. When the Corporation took over the management of the Heath from the GLC in 1989 expenditure on it was £1.2 million. In the following fifteen years £50 million were invested, with the Hampstead Heath Trust Fund contributing a further 17 million, resulting, according to the Corporation, in improved services for Heath users. This coincided with a period of cutbacks on parks and open spaces expenditure by local authorities. For the financial year 2004–2005 the grant from the Corporation to the Heath was just over £5.5 million with an additional contribution of £680,000 from the Hampstead Heath Trust Fund. According to the Corporation this deficit was not, as many suggested, a result of mismanagement of funds. Rather, the budget had risen annually by 3 per cent to cover pay and price increases reflecting, in particular, pay awards and higher National Insurance and pension costs. The transfer from the GLC had also meant that a new department had to be set up to manage the Heath, since under the GLC many of the specialist support services were provided by County Hall. It was not, they were keen to assert, wasting taxpayers' money, since no taxpayer money is spent on its management. At a Hampstead Heath Management Committee meeting on 22 November 2004 various possibilities were identified to balance the budget, including replacing a craftsman gardener with an apprentice, reducing the Conservation Ranger Team, stopping the Heath Diary, Jazz Day and Open Day, reviewing charges for Heath facilities and, importantly, looking at possible options for reducing the £500,000 in lifeguard costs associated with the ponds. The possible closure of the ponds was even mooted, producing an icy response from the ponds' representatives who were present.

It was fascinating that the strength and vociferousness of the response to the Corporation's proposals went far beyond the imaginings of the Corporation, none of whom had visited the ponds, and who clearly thought they were only of interest to a few cranks. Within days campaigners went on the attack. At an extraordinary meeting of the Hampstead Heath Consultative Committee called for 29 November, numerous campaigners raised questions about the smart new vehicles driven around the Heath by rangers and the possibility of raising revenues from the free car parks. What the Corporation had severely underestimated was the constituency with whom they were reckoning. This was not a group of marginalised unemployed with little clout in the public sphere. Far from it. This was a group of professional and articulate middle-class people, many of whom were journalists and public figures, with direct access to the press and legal advice. Many were also ex-political activists from the 1960s and 1970s with considerable experience of how to mount a campaign. According to one report from Valerine Dunn, a member of the Women in Black campaign which supports the Palestinian struggle:

> it was clear that this committee . . . held us in contempt . . . and have no idea of the nature of the ponds and how loved they are by the people who use them, and how they draw in people London wide, from outside London, as well as tourists, so are, in their own terms, an asset to London. . . . Several of the members of the committee sneered openly and spoke against single sex ponds and said there was clearly no place for them nowadays.

By late November the issue had hit the press. On 5 December Eyres (2004) wrote in the *Financial Times* of his incredulity at his discovery that:

> the Mixed Ponds, where men and women, young people and old, of many nations, all colours and creeds, have enjoyed swimming and being close to nature for 150 years, might suddenly and sneakily, in the dead of winter, be closed down . . . when the 2001/2005 Management Plan Policy 83 had stated 'the three natural bathing ponds . . . will be maintained as swimming facilities for the enjoyment of the public'.

By the end of the same week Mayor Ken Livingstone had stepped in to suggest that the ponds should be handed over to a democratically elected accountable body, offering to run the ponds from the Mayor's office – a move dismissed by the Corporation (*CNJ*, 9 December 2004, p. 5). On 8 December more than 100 women supporting the KLPA many of them feminist activists from the 1970s, assembled in a church on Dartmouth Park Hill to voice their concerns and to build a campaign strategy. On the same day Glenda Jackson MP tabled an Early Day Motion in the House of Commons (*CNJ* 9 December 2004, p. 5), while the ex-Labour Prime Minister, Michael Foot, who lives in Hampstead was quoted as saying: 'They are not a fit body to run the Heath and have no understanding of it. They are not responsible people' (ibid). Lord Bragg spoke of the 'free bathing at the ponds as an exemplary tradition of democratic access which has been carefully preserved by generations until now' (ibid). These were voices with clout – and class. By the middle of the month a petition had raised more than 1,600 signatures (*CNJ*, 22 December 2004, p. 3) and a further campaign meeting of the United Swimmers Association had attracted more than 100 people to their meeting in Hampstead's United Reform Church on 19 December (ibid, p. 3). Letters filled the local newspapers week by week, asserting that the ponds were the 'Jewel in the Heath's crown' and 'a sanctuary in the present climate of ever increasing stress and anxiety' (ibid, p. 14), which 'should never be equated with swimming in landed commercial chlorinated pools' (*Ham and High*, 14 January 2005, p. 32) since they represented a site of democratic participation open to all.

At the New Year swim at the women's pond the Chair of the KLPA urged swimmers to turn out in force at the Hampstead Town Hall for the public meeting on 10 January (*CNJ*, 6 January 2005). So too did Sally Taylor, editor of the *Mixed Pond Supporter Newsletter* (No. 1, January 2005) who wittily commented:

> There is an old joke among swimmers. I don't recognise you with your clothes on. So to help you recognise me, I am a tall American woman with long brown hair. . . . When I swim in the pond, I often wear bright pink shower cap (yup, that's me), kind of jarring in that rustic setting but I've found it drives my admirers mad. . . . The biggest problem as you might imagine is that none of the swimming groups appears to trust the management.

The packed meeting, from which 150 people were turned away was declared a sham (*CNJ*, 13 January 2005, pp. 3–4) – 'an exercise in pro forma futility' – by those who attended, as speaker after speaker condemned the proposals. On 14 January the local newspaper (*Ham and High*, p. 1) revealed a secret plan for the approval of a £100,000 contingency fund to cover the cost of defending the Judicial Review on the banned unregulated swimming, so adding further fuel to the fire. As a member of the United Swimmers Association, Peter Cumming, put it: 'The barrels are not empty. If they have a contingency for legal procedures, then they should have contingencies for other things' (ibid). Expressing the view held by many of the campaigners, on its front page the paper declared: 'the threat to swimming on the Heath ponds can be attributed to only one thing: a monumental failure of management'. And, as another speaker put it, 'If the CoL abandoned its social responsibility of maintaining free facilities on the Heath it will certainly be inviting civil disobedience on a large scale, and fighting it could be very expensive.'

The letters page of the *Camden New Journal* (20 January 2005, p. 18) revealed the severe state of impasse and personal animosity reached. In many of the letters the three key players from the Corporation of London, Catherine McGuinness, Chair of the Hampstead Heath Management Committee, Superintendent of the Heath Simon Lee and Jenny Adams, Director of Open Spaces, were singled out for rebuke. They were variously referred to as forfeiting public confidence and inspiring mistrust, and as the Corporation's stooges. The only solution, letter writers suggested, was for the trio and the Corporation to think again, admit their mistake and back down. What was needed was a strategic evaluation of the Heath's administration costs or, as some authors put it, 'shocking escalating costs' and mismanagement of funds. The Chair's letter in response to what she saw as previous misrepresentations of the Corporation's position asserted that only 46 people were turned away, not the 150 proclaimed by the campaigners. Her letter was placed prominently in an outlined box on the letters page. The following week (*CNJ*, 27 January 2005, p. 15) the letters page was again filled with aggrieved voices. Various writers testified to the large numbers turned away, arguing that the Corporation's figures exclude all those 'who did not fill in the consultation form as per the CoL's request, the obvious reason for this is their belief that the CoL would pay lip service to their views'. The Chair of the United Swimmers Association, 'in the spirit of Athenian democracy now abroad in Hampstead', took issue with the Chairman of the Hampstead Heath Society's accusation that the campaigners were extremists:

> the extremism is on the other side. . . . Is seeking justice extremism? Is appeal to law seeking lawful redress political gain or the act and expression of public spirited desperation? . . . All I desire is a hook on which to hang my clothes, a natural pond and a cold shower.

And another railed that 'if the Corporation had their way the whole of the English coastline would become a sanitised commercialised theme park and we would have to pay to go for a swim in the sea'.

The following week the heat had not abated (*CNJ*, 3 February 2005, p. 2, 15) with an article calling once more for greater transparency for the Heath's £5.5 million budget and better management of the funds, and more letters expressing fury at the CoL's lies, inflated salaries and relentless drive to charge for this natural and public facility although the terms of the Hampstead Heath Act of 1871 protect the Heath for all users. On 22 January a further meeting, attended by hundreds of supporters, was organised at the local Everyman Cinema at which the film *Swimmer* was shown and Roger Deakin, author of *Waterlog* (2000; an account of his journey round England swimming in unorthodox sites), spoke on behalf of the campaign. In the week that followed, the London Mayor, Ken Livingstone, 'waded' into the row, describing the Corporation's proposals to charge swimmers £2 per day (£1 concessionary rate) to use the ponds as 'a silly idea' (*Ham and High*, 11 February 2005, p. 1). If the Corporation was unable to honour its promise, made when it took over the ponds, to keep them open and free, they should be transferred, according to Ken, to the Greater London Authority.

The date 22 February 2005 marks the dénouement of the story narrated here. At a meeting of the Hampstead Heath Management Committee at the Livery Hall of the Guildhall, more than 100 pond users, including swimmers from Russia, gathered to hear the Committee debate proposals to cover the shortfall of £150,000 in the budget: car parking charges to raise £50,000, £20,000 of savings from cutting opening hours, and charges for the ponds to raise £80,000. A further option was an annual swimming pass of £100. A petition with 6,884 signatures was handed to the Corporation, including 50 signatures from the Walrus Club in St Petersburg. As the Committee reiterated its commitment to the proposals for charging, speaking with emotion of the vilification of the Heath bosses during the campaign, members of the audience shuffled and sighed in their seats, with occasional shouts of 'Shame on you', 'Sack the management' and 'Why have the funds been mismanaged?' as exasperation rose in the hall (*Ham and High*, 25 February 2005, p. 10; *CNJ*, 24 February 2005, p. 5; personal observation). The proposals were passed with a minor concession to the swimmers whereby the proposal for cuts to the swimming hours at the mixed pond was withdrawn: a

Plate 4.4 A member of the Save the Ponds campaign with a petition of support from Russian swimmers.

whimper rather than a bang. But time had been bought and the swimmers vowed to keep fighting to keep the ponds free of charge.

The end of the saga as this text was in its final stages of preparation in the summer of 2005 reflected the same feisty and humorous resistance that had characterised the ponds campaigners throughout. In early June the Corporation of London placed machines that looked much like pay and display parking machines at the entrances to the ponds, requesting swimmers to pay £2 or £1 concession. An article in the local paper with a characteristically humorous headline – 'Pond swimmers make waves over charges' – reported the outrage and dismay at their insertion: 'the bathing ponds used to be one of the few places where money didn't matter' and 'I think it is the thin end of the wedge' (*Ham and High*, 17 June 2005, p. 6). Sitting in the changing rooms some days later I listened to the conversations amongst bathers, none of whom had paid, some of whom had not even noticed the machine, and all of whom considered it an absurd intervention into what they saw as their public space. Voices of incredulity, dissent and disgust were to be heard all around.

CONCLUSION

The relationship between the move to curtail the winter swimming and the expenditure cuts was unclear. Some campaigners claimed that the litigation around the winter swimming campaign was muddying the waters, while legal experts on the case declared there was no relation. As early as February 2004 one of the winter swimmers (BBC producer Piers Plowright) pointed out that cuts in expenditure had motivated the moves to curtail early morning swimming at the ponds: 'These cuts are mean. Hampstead Heath has been free to everyone for a hundred years. But if economics have become the priority, why don't we turn the whole area into a theme park?' (*CNJ*, 13 February 2004, p. 6). It seems probable that the two initiatives were not directly connected – at least initially, despite the financial implications of lifeguards deployed at the ponds for morning swimming. However, it remained in the interests of the Corporation of London to foster the confusion throughout the whole process, since it provoked internal tensions, at various junctures causing some pond campaigners to put pressure on the early morning swimmers to withdraw their case.

This interconnected narrative reveals a number of things. On the one hand it reflects state practices designed to cut expenditure and minimise and manage risk in the context of a wider culture where the regulation of risk has begun to replace the state's role as direct employer or provider of services (Hood *et al.* 2004: 4). On the other we see urban subjects reshaping and contesting the citizenship games (in Rose's (2000) terms) which govern them. In the context of new forms of government where there is a moral responsibility for citizens to take on risk reduction for themselves in their own localities through the policing of conduct (neighbourhood watch, zero tolerance, etc.), these individuals were asserting the desire to manage their own risk in this space too. This was, as we saw, a mainly educated – politically, culturally and socially – group who were thus powerful enough to resist what they saw as an intolerable incursion on individual liberty and, perhaps more importantly, on serendipitous and pleasurable embodied sociality in marginal spaces of the city. This campaign thus represented successful resistance to forms of government, so abhorred in much contemporary socio-political debate, which are articulating new modes of personal and public responsibility in the everyday spaces of cities. What was defended was the right of people to occupy and rub along in public space that was in danger of being taken away by over-zealous regulators. This diminution of unregulated (or informally regulated) public space and its replacement with formal regulation and prohibition, I suggest, produces the dead and abandoned spaces that Richard Sennett (1974, 1996) and others have railed against, while an exaggerated concern with risk and protection erodes public life.

5 Disrobing in public

Embodied differences in bathing sites

Public spaces are produced by the people who connect, disconnect, flow through and transect them, and by the people who play, laugh, cry and interact in them – in other words they are spaces of desire and affect. Encounters constitute the very spaces within which they are enacted; encounters bring a place into being. In this sense we can think of urban encounters as choreography (Pile and Thrift 2000). Public spaces are performed as well as being spaces of performance. The very specificity of the city, its shape and form, the way it is designed and managed, makes a difference to the bodies that are therein produced. Grosz (1992: 301) puts it this way: 'the form, structure, and norms of the city seep into and effect all the other elements that go into the constitution of corporeality and/as subjectivity. It effects the way subjects see others.'

This chapter seeks to interrogate the common and contrasting ways in which particular bodies are produced by, and themselves produce, public bathing spaces in the city. This mutual constitution of space and bodies, and the ways in which this is represented, contested, regulated and lived is explored through two kinds of sites: first, the municipal Turkish baths in Britain; second, the natural swimming ponds of Hampstead Heath in North London make another appearance to tell a different story. My particular interest is how in different historical periods the discursive practices and strategies deployed by the people advocating, constructing and managing these public bathing spaces have worked to define bodies in distinct ways. Just as the expectations of those controlling these spaces have changed over time, so too have the bathing practices, performances, subjectivities and resistances of those using these spaces. What I want to explore here also are the gendered, sexed, classed and racialised subjects produced in these spaces and, in particular, the ways in which practices of bathing have constituted, and been constituted by, dominant norms of contemporary femininity and masculinity. I hope through discussion of this site to expose how difference has been played out in specifically embodied ways, and how these sexed and gendered cultural practices have been represented and regulated publicly.

How spatial practices have produced different bodies has been of interest to geographers for some time. Drawing on Butler (1993, 1997) we have the notion of bodies as discursive, produced through language, where social identities are performed, and where sexed bodies, gender and heterosexuality are constructed as a social relation performatively and discursively produced rather than existing prior to discourse. These processes of the social construction of sex and gender are firmly embedded in power. As Gregson and Rose argue (2000: 434), performance – what individual subjects do, say, 'act out' – and performativity – the 'citational practices which reproduce and/or subvert discourse and which enable and discipline subjects and their performance – are intrinsically connected, through the saturation of performers with power'. Moreover, performances are enacted in spaces and thus bring these places into being, places where performers are produced by power and the spaces are themselves performative of power relations (ibid: 441). Such an approach points to the instability of identities and social differences, and the significance of looking at their construction in particular sites of the city. Following Grosz (1995: 108) also, there is the notion of bodies and cities producing each other in a multidirectional network of flows where each defines and establishes the other:

> It is not a holistic view, one that stresses the unity and integration of the city and body, their 'ecological balance'. Instead I am suggesting a fundamentally disunified series of systems and interconnections, a series of disparate flows, energies, entities, and spaces, brought together or drawn apart in more or less temporary alignments.

Thus, I am interested here in what kinds of bodies, sexed and sexualised, are produced in public bathing sites, and how bodies in turn produce these spaces in particular ways at particular times. But this too, I argue, happens in a political/social context – something Grosz and Butler are less keen to explore: in particular, forms of regulation and governance allow and make possible different kinds of bodies and performativities to occur. In the Victorian period, what we see is a shift from discursive strategies of cleanliness mobilised by often self-made philanthropists and experts on behalf of the working classes – sometimes segregating them, sometimes hoping to include them within discourses of egalitarianism and public space. Intrinsic also to these discourses are preconceived notions of sex/gender. What we see in the current period are new discourses of risk defining new strategies of regulation of bodies on the one hand, as was explored in the previous chapter, and the forces of privatisation segregating rich and poor bodies in new spaces on the other.

STEAMING BODIES – TURKISH BATHS AND THE PERFORMANCE OF DIFFERENCE

I want to turn to municipal Turkish baths as a fascinating site of disrobing in public both to explore how, historically and today, they embody notions of living with difference, and to consider how sexed/gendered and racialised bodies have been produced in these enclosed, but also public, bathing spaces. At the turn of the nineteenth–twentieth century there were approximately 600 Turkish baths in Britain. They suffered a steady decline over the following 100 years, with only 21 remaining today (Curry 2002). But there are signs of a revival as the Victoria

Plate 5.1 The interior of Porchester Road Baths, Bayswater, London.

Baths in Manchester has been marked out for restoration having won the public's vote (282,018) on BBC2's *Restoration* programme (*BBC News*, 14 September 2003) and campaigns have been mounted to preserve some of the remaining baths. In many countries, particularly Morocco and Turkey where they originated, but also in Japan, Korea, Russia and countries of Eastern Europe, the Turkish bath lives on as a vibrant and enthralling secret world (Brue 2003). In Northern Italy the old Roman baths in Bormio have been restored and revamped to provide a luxurious indulgence for tourists and rich weekenders from Milan. In Britain, though, the old municipal Turkish baths are not sites of exclusion and wealth; rather, they retain a dilapidated, mundane and egalitarian ambience, often hidden away in old buildings, but accessible and affordable to those aware of their existence.

A website dedicated to the Victorian bath (www.victorianturkishbath.org) which provided much of the historical information presented here gives an extensive account of the origins of the Turkish bath in Britain. This is a type of bath in which the bather sweats in a room which is heated by hot dry air (bathers progress through a series of increasingly hot rooms until they sweat profusely). This distinguishes the Turkish bath from the medicated vapour bath, or the steam baths usually known as Russian baths, which existed before 1856 (Shifrin 2004).

In the mid-nineteenth century a Member of Parliament, David Urquhart, with an extensive interest in Russia and Turkey, galvanised groups of working men into Foreign Affairs Committees, which later became the nuclei of the Turkish Bath Movement, to promulgate and campaign for his views on foreign policy. These same working men were responsible for the first Turkish baths to be built in England. For Urquhart, himself a neuralgia sufferer, the Turkish bath not only exhibited remarkable therapeutic properties but was also a means of promoting personal hygiene and cleanliness. Most significantly he wanted Turkish baths to be public places available to the poorest members of society, offering a space where people could easily mix together and thus potentially breaking down class barriers. Under his direction and with his support (sometimes financial), over 35 Turkish baths were set up by Foreign Affairs Committees or by one or more of their members in the main towns of Britain. Visits to the Turkish baths became a regular occurrence in everyday life for many city dwellers. Thus, for example, there is a record of James Whistler, the American artist, writing in 1865 to his patron, Lucas Ionides, suggesting an outing to the baths before dinner: 'I shall go to the Turkish Baths in Jermyn Street at about 4.30 or 5.0 You had better come there and join me. We will then come back together. If Aleco [Ionides] will come bring him with you' (Boyd 2003).

When David Urquhart died in 1877, *The Times* wrote: 'Whatever may be thought of his political idées fixes, he has, at least conferred one great boon upon England in the introduction among us of the Turkish bath, the one Turkish institution which it is certainly desirable to adopt' (quoted in Shifrin 2004).

This was a period when cleanliness and hygiene underpinned health and social policy initiatives, particularly those directed at the urban poor. At the time of the Wash Houses Acts in the mid-nineteenth century, which instituted public baths for the poor, there was no mention of Turkish baths because they did not yet exist in Britain. As a result solicitors in many local authorities advised that they could not legally be included in bath houses built at public expense, unlike vapour baths which had existed prior to the Acts, and were therefore permitted (ibid).

This setback did not, however, deter Urquhart who started to work with Richard Baxter (an Irish doctor and hydropathist who had successfully established St Anne's Hill Hydropathic Establishment near Blarney in Co. Cork) on experiments with new methods of building Turkish baths. By the end of the century at least 370 baths were opened as a result of Urquhart and his Foreign Affairs Committees' campaign, though his dream of egalitarian access to these sites for the working classes – a notion of producing de-classed bodies – was constrained by the profit motive of the private establishments. The first council-run Turkish bath was opened in Camberwell in 1905. In their attempts to attract people from all classes advertisements for Turkish baths emphasised that everyone would have access to the same privileges, with distinctions being drawn by a pricing structure which attached different prices to different times of day. As Shifrin (2004) points out, this separated bathers no less than if they had advertised separate classes of bath.

Not only were class differences produced in these sites, so too were gender differences. Of the 370 baths that were opened during Victoria's reign just over 100 catered for women. Shifrin (2004) quotes an anonymous leader in the *Manchester Critic* in 1872 who advised:

> If ladies only knew what a real and lasting beautifier the Turkish Bath is, they would abandon all agitation for 'Women's Rights,' &c, and at once build a Turkish Bath for their own special use. By using it, they would thereby render themselves so fascinating and beautiful, that there would be no resisting any appeals they might make to the weaker sex.

Turkish baths for women grew increasingly popular. By the late 1880s, advertisements, like that of the Pilgrim Street Turkish baths in Newcastle upon Tyne in the late 1880s which portrayed Lillie Langtry, extolled the virtues of baths with the promise of softer skin and beautiful hair. Women, embedded as they were in daily domesticity, were also enlisted to promote Turkish baths to their men:

> To show the delightful influence of the Turkish Bath, suppose a man comes home ill-natured, jaded, and weary with the affairs of the world, cross with his wife, and quarrelling even with his dinner; let the good wives who hear me take my advice, and tell their husbands to go wash in a Turkish Bath, and they will throw off their ill-humours – mental and bodily – and then return delighted, as I am sure they will be, in spirit, happy with their wives, contented with their dinners, and playful with their children.
>
> (quoted in Shifrin 2004)

By the late nineteenth century 104 baths were open to women, 38 of which provided separate facilities in the same building with separate entrances, and 64 of which provided separate sessions for women. Advertising material emphasised that women had no need to fear intrusion and their privacy was respected. As we see too in the same period, there are descriptions of a feminised domestic space. For example, the baths at Brighton had a large cooling area, elaborately decorated and furnished, where 'feminine taste and elegance of disposition of being was, of course, considered and provided for' (Shifrin 2004).

Plate 5.2 A former site of municipal baths, Hackney, London, now a Vietnamese cultural centre.

During the twentieth century there was a steady decline in the number of Turkish baths in Britain. The Turkish baths in Harrogate, a northern spa town, are a typical case. Built in 1897 in a flamboyant Moorish design with great Islamic arches and screens, walls of bright glazed brickwork and arabesque-painted ceilings, in their heyday the baths were frequented by royalty, politicians and latterly film makers. According to the Harrogate baths website:

> In 1897 Princess Alix Hesse and her sister, Princess Victoria of Battenburg, came and amused themselves by racing their bath chairs through the streets of Harrogate, and it was said, that on any morning during the spa season, it was possible to hold a Cabinet Meeting in the Pump Room, so many ministers visited the great treatment centre.

In 1969 the treatment centre was closed down, according to the site, due to the emergence of modern treatments and technology, though the Turkish bath was maintained as a functional shrine to this forgotten era (www.harrrogate.gov.uk/turkishbaths/history.html). Less fortunate was the Brighton Hammam, established in 1868 and converted into the Academy Cinema as early as 1910 (Shifrin 2004).

By the early twenty-first century, only 21 of these Victorian Turkish baths remained, as they had been decimated by a combination of the massive expansion in private baths, cuts in public expenditure on local municipal facilities and the growth in private leisure facilities and clubs – many of which have saunas, jacuzzis and steam baths. The irony is that a well-managed preservation or restoration of these baths could have been a good source of income for local authorities as body pampering and beauty treatments have expanded since the mid-1980s – reflected in the massive growth of consumption on body products in specialist shops like the Body Shop and Lush and in high street chemists. In the early 2000s there have been signs of a renewed interest in Turkish baths in Britain, evident not simply in Manchester but in the success of the campaign in February 2004 to save the York Hall Baths in Bethnal Green, East London, from the threat of demolition. Travel books like the Lonely Planet and Rough Guide series also frequently refer to the pleasures of the Turkish baths for visiting tourists.

Turkish baths produce different bodies which themselves perform cultural practices refiguring these spaces differently across time (since the same baths are alternately used by men and women on a daily basis). Striking in my visits to the women's baths (a male researcher visited the men's baths for me, see Appendix) were practices of living with embodied differences without tension. A comparison of the baths on men's and women's days revealed striking contrasts, with a construction of domesticity on the women's days and a sexualisation of men's bodies during men's sessions. This follows a long tradition whereby the baths functioned as one of the sites of gay male sex when sexuality and intimacy between men could only be expressed in private (Chauncey 1994). In each of these spaces, temporally differentiated, public–private boundaries were confounded and disrupted, with sexual and social practices more usually associated with private life

Plate 5.3 York Hall Baths, Bethnal Green, London, opened by the Duke and Duchess of York in 1929.

Plate 5.4 Entrance hall to the baths: no money was spared in these early days.

imported into the baths and creating new private/publics and new meanings of spatially appropriate embodied practices. Striking also was an easier acceptance by women of the idea of gay and heterosexual women sharing the space, and a more open expression of homophobic feelings by the heterosexual men who visited.

Let us enter the baths today. An interview with the attendant for the women's sessions in a Turkish bath in Central London captures a sense of the place on women's days. These were sites of vibrant sociality across race and ethnicities, as well as age, sexualities and class. Palpable was this woman's pride in a little kitchen she had established in the TV room from which she sold tea, toast and sandwiches, constructing this site as social and domestic:

> Got my regulars on Sundays – once a fortnight most of them . . . got a nurse . . . got Kim – she lives in Buckingham . . . I think she is a call operator . . . they have all got friends been coming for years n'll . . . she's about 40 . . . she's single . . . she's English . . . she enjoys coming here, she made it sound so good for the gals she works with she ended up bringing them but she likes coming on her own so she didn't do herself any favours – fortunately one has moved and the other one isn't coming back till October – so she has some peace again. Got Rona the nurse she does the night shift . . . she comes in Sunday morning – she is 52, she is single . . . Scottish. She comes every week . . . she's become a very good friend . . . on Sunday afternoon it is hilarious out here, I have to tell them to be quiet cos of that lot in there – another couple of women who work at the local hospital . . . they come too . . . in their forties – they come on Fridays, they swap it around with their shifts. They all get friendly since I came here. . . . Coloured ladies come here too – two on a Wednesday – one is from Africa – the other one was born in this country but don't know what she is, she's very dark – the other one is mixed race . . . I know that Brenda was at university but had to pull out cos of her family situation. I know she is some kind of preacher . . . they talk to others too. They have started coming out here – though they don't smoke – they can hear the laughter so they've joined in now. Wednesday we got two Turkish ladies and my Italian lady Anna and they come in every week – the two Turkish women have been coming for years – they are in their 70s . . . they didn't used to speak much English, but now I carry their bags down for them and carry them up – just being helpful, we've got talking more. The Italian woman – she is regular – I call her Lady Face ache – been coming for donkey's years – and the staff upstairs call her Donkey's Ears. She comes out and says the steam is too hot . . . I've just settled myself down . . . and she comes out and says the steam is too cold . . . and I have to go all the way upstairs again and turn it down again . . . and then she is in the hot room all dressed up in black bin liners – then she comes out and says Anna it is too noisy and I try and explain we are all going to go deaf some day . . . she is in her 60s . . . she brings all her own special foods – it was what she calls her detoxing day.

Plate 5.5 York Hall, Baths, Bethnal Green, London.

> . . . They are all nice in their own way . . . Sundays – usually get people who work Monday to Friday . . . you get policewomen down here and that . . . all walks of life . . . we talk about one another's families – I got it right down here.

At another Turkish bath in East London – an old and dilapidated building constructed in the late nineteenth century – I fell upon one of the most multicultural sites I have ever encountered in London: women from Morocco, Ghana and various other black African countries, Latin American, Afro-Caribbean, Chinese and very few white women. The women wandered around with buckets full of unguents and oils, with their hair wrapped in colourful materials, and the place was filled with a cacophony of sound, laughter, chat and screams of delight. Sitting on black plastic bin bags on old wooden benches surrounded by dripping, peeling walls, these women inhabited a space where race/ethnicity seemed to be celebrated and lived. As one woman said, 'Well, look, we are all the same when you get down to it'. Many women were rubbing each other with mittens and cloths in an unself-conscious way. A Moroccan Muslim woman talked of meeting Jewish women there, another woman lent me her beautiful oils laced with orange blossom for a massage by a Chinese woman who gave privately arranged massages on the benches in the middle of the locker room in full view of all who went past. An extraordinary place.

Plate 5.6 The now dilapidated interior of York Hall Baths.

But here on the men's days was another world, where many of the men were white working-class cabbies, where the non-whites – mainly Afro-Caribbean-were discursively reduced to talk of penis size and appearance. Notable in this interview was the fantasised, and desired, figure of the large African penis:

> Sitting over there, there were a few gentlemen and an African and very old gentleman were sitting, having a conversation for about an hour, they must have known each other maybe, being regulars, and then there was another gentleman, sitting over there, he was reading the newspapers for about an hour, flashing out, good, he liked himself, because all the people were perving over him and he was just going from room to room saying, look at this giant cock and you cannot have any of it [laughs] . . . bloody pig.

In this space, where men came to partake in gay practices, the fears of some straight men who landed there unknowingly were evident. Perhaps this represented men's fear of their own homosexual feelings and desires which are projected outwards into disgust. Sibley (1995), drawing on Kristeva's notion of the abject, explores the compulsion to exclude threatening others from one's proximity. It is connected, he suggests, to anxieties over maternal loss, propelling individuals to maintain the integrity of self. 'These urges to prevent boundary violation feed on stereotypical images of repulsion mapped on to particular social groups. The potential for abjection is thus present when spatial orders are called into question, blurring the distinction of pure and polluted' (Hubbard 2002: 370–371). Or perhaps they had been influenced by media representations of gay men as hedonistic and sex obsessed, their visibility equated with the imagined provocation of hetero-sexuals (Sibalis 2004). One interview vividly captures the ambivalent feelings – of fascination and disgust – produced in this straight man:

> I felt very uncomfortable, because I could tell there was big homosexual activity going on in there, and for a heterosexual guy to go in there, it is quite intimidating, there is a lot of checking you out. . . . There are people playing with their dicks in there, stroking each other. There was one bloke massaging . . . the . . . I could tell he was not massaging the usual regions, em, I felt really intimidated, I did not feel comfortable.
>
> It is just when I walked in the changing room, I did not clock straight away, I was being my usual loud self, the men were just looking at me, a lot of men walked around naked, you get a couple of them but not that many and then that black guy walked in, king don, whistling and then I went to the steam room, that's where a bloke started playing with himself and the guy sitting next to me, I spoke to him briefly and then he kept looking around at me, looking down on my towel and hope he is going to rise up and shit like this, I was . . . like sweating and you can't help it to touch your body and get rid of the sweat and probably think oh he is turned on here, he likes it, and I am like fucking skinny with a goatee; oh, he is a new fish here, that's what they thought.
>
> . . . I go to the sauna, it was so small and everyone sitting so close, I thought there is something dodgy here and then they were massaging each other, I thought me . . . this place is fucking . . . I am not going to fit here and then I felt it was really uncomfortable cause it was too hot there, I started to think, I am going to have a panic attack here, I felt claustrophobic and all the

eyes were on me, blokes are just sitting opposite me, just checking me out, so I try to sit more manly, but that made me more fucking . . . you know just sort of pushed my shoulders back and exert presence, I think that made me maybe more attractive to them, so I sat cross legged and at one point I have realised that my towel has risen up so you can see my bollocks, I thought shit this is getting worse and worse and of course when you try to wipe the sweat off I was pushing my hair back with my hands and that would be seen as quite sexual I suppose and I was thinking this is going pear-shaped and then you think oh I am just going out for a breather and sit on the bench just outside the steam room and that's even worse, they sit next to you and start looking at you, they see you cooling off, they make conversation: I got changed, spent time in the steam room, I went about four times, had a couple of showers, cold showers, just hung around on the benches just outside the steam room and that's all about it. I did not go to the sauna, I did not feel comfortable, every time I looked at the sauna, there was only one person in there and just felt it was a bit dodgy and if you would walk past that person and they would give you the eye like come in with me and physically lots of them were big, big built, I was probably one of the smallest, very hairy, shaved downstairs department, definitely, there was lots of them, it seemed to be a lot of interaction between the younger people and the older people with the massaging, quite a few times. In the steam room you go in and basically there would be a younger chap lying down getting massaged with body oil by one of the older chaps and the massaging of the arse was very popular . . . I just did not enjoy the whole experience, I just really could not stand it.

But other men were more sanguine and the Turkish baths themselves varied, from places like Swindon which is largely white and middle class, to other baths in London which had a much more diverse population, where men got along easily, enjoyed an easy sociality, and even, if heterosexual, appeared little perturbed by the gay life they saw there. At most there was an expression of awkwardness rather than hostility:

I felt awkward upstairs when a very muscular man (ex-RAF) spread his towels in front of me and started applying talc powder to his body (especially his groin). To me this is a public room and that, he should have done it in a private room. I can understand when people want to get changed, they take off their body wraps/sarong and dry themselves off and wear their clothes, but when people indulge in such a behaviour, I felt very awkward.

While for another:

funny but I would not say awkward, seeing a couple of the old guys just, with all their gear off and just walking around and they did not care about it and that to me, I don't think it was awkward, it is just funny.

For many men visiting the Turkish baths meant relaxation, chatting with others – usually on topics classically designated as masculine like football, or on health and family issues. For example, in another Turkish bath:

> I mean the atmosphere was very good, I felt safe all the time – I did not feel threatened, I felt fine. A lot of people just sitting and chilling: exhibitionists there, some lurkers, just follow you around and stuff like that, but I mean it was all good.
>
> The seating area opposite the scrub room also has a social feel to it. With the exception of two black males, all the other men were white. The conversation was quite animated, they were talking about football. It has the same feel as the pub on a Saturday afternoon. They all seem to know each other.
>
> You have a range of ages in there, there was quite a few old people there relaxing and they tended to be quite free, walking around, some naked and others had their towels on, people very very relaxed. To be honest, I would go as far as to say that there was business transactions going around in the place, be that legal or illegal but not in the way that you feel threatened, everyone is relaxing and it is non-intrusive, no one is looking at you, at anybody, just continuing their business with themselves or whoever they've gone with.
>
> (black male, age 32)

Interestingly, it is the work sphere, not the domestic, that is imported here, as men, in this respect, perform their public, not their private, selves in these sites.

Another man described his affection for one of the London baths which remains:

> It feels that it's got an old feel to it – it is like stepping back into time. It is like you step out of the haste and bustle of London living and then you step into this place, which is really going at a slow pace, it has its rules dictated by its parameters . . . like everyone does their own thing when they go there, everyone goes there for their own personal reasons as opposed to you having to catch a train, you all get into that train to go to that place at a particular time. There was no pressure in there, you do your own thing, you interact with people but other than that . . . you would not know about it by just walking on the high road, that's what I like about it. It's got a secretive element to the whole place. If you walk past a bar, you say that's a bar, I am gonna go in there and check it out and see what everyone does in there, whereas this place, it is like oh, it's a place where you go or you don't go: I am just going into the steam room, and I am gonna there at this particular time, and there is no checking out element, you go there for whatever reasons [laughs], nice, relaxing whatever. . . . In that place, it is like time-out, your own personal reasons for going in there, relaxing a bit, do a little of business, be it to meet someone you like [laughs], so it is an amazing place. Lovely.

From my own research it did indeed appear that these Turkish baths were democratic spaces of sociality and connection, offering opportunities for contact across class, race and other differences. But whether they offer more of these possibilities for gay men than do some more commodified and marketised spaces of gay culture is hard to assess. Challenging what they see as possibly romanticised accounts (e.g. Rofes 2001; Delany 1999) of these kinds of sites, Bell and Binnie (2004: 1812) are certainly sceptical.

HAMPSTEAD PONDS

In the last chapter I gave an account of the Hampstead ponds and state regulation. These same ponds reappear here to tell a different story, one which bears many resemblances to the previous account of the Turkish baths. What re-emerges is the appropriation by gay men of what is simply a male-defined space and, in the women's pond, the production of a female culture embedded in distinct customs and practices, where a positive camaraderie seems to have defined the space – at least for the regular users. This gendering of the space in the women's pond is once more marked by practices of importing the domestic via food, the nurturing of others and even the presence of a pet cat, which – like the baths – appears to confound the boundary between public and domestic space in intriguing ways that are virtually absent from the men's pond. In Butler's (1993) terms, practices of femininity are iteratively imitated and performed, constructing the notion of the women's pond as highly gendered and different from the men's pond 500 yards away.

The women's pond today attracts a variety of women including Orthodox Jewish women.

> Hasidic Jews – the ones who wear the wigs who are not allowed to be seen bathing – they come here – put their hats on – soft ones – they are great – very nice women . . . sometimes come on their own and sometimes in a group. . . . They have their own particular relationship with the pond. . . . It is fantastic.
>
> (Chief lifeguard, June 2003)

Interestingly, according to the lifeguard quoted here, fewer Muslim women, who also require single-sex bathing, use the pond, though she had no explanation why. The women come from every class background and occupational group from opera singers to nurses, though the winter swimmers are more of a distinct class group (middle to upper middle class). There appear to be proportionately fewer black women than in the London population as a whole, particularly among the older women. Women talked of coming to the pond for years, loving its peace and tranquillity, the possibility of sunbathing topless on the grass surrounded by trees in the middle of a city, reading, talking, swimming and connecting with others. Gwynne, who was 92 years old, had cycled from Camden to the pond every day of the year for 50 years. On her ninetieth birthday she arrived for her daily swim to find that the other all-year-round swimmers and lifeguards had arranged a surprise party for her, having adorned the trees with balloons and baked a large birthday cake. 'I was overwhelmed,' she told me. A common discourse was: 'It doesn't matter what you look like, whether you are fat or thin or young or old.' Here women of diverse identities and subjectivities are free to perform their differences and literally construct their bodies as women with tattoos and piercings lie side by side with women who twenty years ago would have observed such markings with incredulity.

This space has seen a shift in users and cultural practices over the last 20–30 years. Lifeguards and women described a greater cosmopolitanism as the pond has found its way onto tourist websites or into travel guides. Globalisation has reconstituted even this unique local space. Others mentioned a visible presence of lesbian women since the 1970s. Though there had been some tensions in this early period of 'coming out' politics, where many women claimed a public lesbian identity and practice as a militant political gesture, the current chief lifeguard claimed that tensions around sexual identity had eased off since the 1990s in line with legal changes and a more liberal social/sexual environment. As one woman in her sixties put it: 'I notice more women canoodling – but it doesn't bother me.' She, and a friend who was sitting with her, recounted stories of men they knew:

> who were unable to handle the men's pond now. . . . They just go for a swim and come straight out to sit on the bank . . . it is the macho thing isn't it – makes them feel threatened. Not how men are supposed to be is it? Women don't care – it is a bit of a laugh really.

Yet according to the lifeguard:

> we still do get complaints about lesbian couples from time to time – private things in public really . . . the other day two women complained . . . they were older, they felt uncomfortable. But they said they would feel just the same if it was a man and a woman.

Such complaints are dealt with in a very low-key way. This relatively easy attitude to gay people contrasted strongly with the men's pond. There were nevertheless occasional complaints from women who found such behaviour distasteful and these are usually resolved through a quiet word from the lifeguards. The ladies' pond, then, is a place of co-mingling of strangers, of embodied difference performed in a public space of the city with apparently little tension.

This sociality finds its public voice in the Kenwood Ladies' Pond Association (KLPA) – a democratically elected group which represents the interests of women swimmers, providing newsletters and arranging events. It has produced a cook-book, *The Hearty Swimmer* (2002), with recipes and tips for the winter swimmers. The fierce identification of its members with this space is graphically illustrated in a dispute over a public art installation, which itself throws up important questions about who has the right to represent and monumentalise public space. It is also a

Plate 5.7 Long-term swimmers of the women's pond enjoying the sun.

story of imaginary ownership. In July 2000 the artist Orla Barry installed her *Across an Open Space* as part of a series of public art projects, the North London Link, co-ordinated by the Camden Arts Centre. According to her website:

> Much of Orla Barry's photographic, video, performance, text and sound installation work searches for the place where myth, memory and a robust and sensual physical reality intersect. . . . She inserts her distinctive blather of words into everyday talk. By using several forms of address . . . Barry invents a fiction of multiple 'I's, which enrich our understanding of the unfixed, multiple nature of identity.

In this instance, Barry placed signs around the ladies' pond, and other sites on the Heath, which played on the official ones – 'No dogs', 'No sunbathing', 'No cycling', etc – with text like 'Rootless we stand', 'A flimsy feeling of happiness', 'Looking is wanting'. The project was developed from over a year's rambling on the Heath, observing the minutiae of everyday life and listening to snippets of conversation. In conjunction with her signs, and drawing on a story-telling tradition (Smithson 2000: 29), she published a magazine called *Refuges* which played on the Heath's place as a refuge from frantic contemporary urban life. Barry encouraged local people to contribute to, and be involved with, the project. At an early stage in the process she met with several of the members of the KLPA, proposing to take photographs of the winter swimmers and their glowing bodies as they came out of the water (*Hampstead and Highgate Express* (hereafter *Ham and High*), 4 August 2000). The proposal to exhibit these in the changing room for the women to see was discussed and accepted by the committee. Barry came to take the photos and the women heard no more. As the Chair of the Association put it: 'Then suddenly, on July 21 we were appalled to be confronted by large boards of a virulent green, bearing inexplicable words. There had been no warning that this was going to happen, nor any explanation that this was "art"'.

The outrage felt by some of the swimmers, who were never identified, was such that they felt compelled to vandalise and remove some of the signs. The heated war of words which erupted in the local press following the event led to accusations that the art had been destroyed by fascists and vandals. The police treated it as an act of criminal damage. Orla Barry admitted being stunned by the reaction (Etoe 2000: 4):

> That a handful of people would decide for the thousands of others who use the ponds and the Lido every summer, that my temporary texts were not suitable and start to vandalise and remove them is very curious . . . I think my project has raised a lot of questions about 'public space' and people's claims to it.

Indeed. But for the users of the pond it was a different matter. A petition to remove the signs was rapidly drawn up and signed by 63 swimmers, and various writers to the press expressed their dismay: 'The women who swim there may not own the pond but they care for its natural beauty . . . and protect it jealously. We suspect our feelings run deeper than Orla Barry's,' said Margaret Hepburn (*Camden New Journal* (hereafter *CNJ*), 27 July 2000). Another wrote: If . . . [she] thinks of the heath as a place of refuge why did she violate ours? Even bringing men in to set up the notices showed insensitivity in my view' (*CNJ*, 10 August 2000).

No such battle around representation has occurred in the men's pond and no organisation like the KLPA exists, only the Lifebuoys which is 'a cliquey club of old men, which has managed to establish certain privileges like playing shuttle-cock' (lifeguard, women's pond). (There is the United Swimmers Association which purportedly represents the interests of both ponds but in fact was set up by three people and is undemocratic.) The men's pond is a distinct space, physically and culturally, producing different sexed and gendered bodies and practices. A lifeguard at the women's pond put it this way:

> it's like moving from Norfolk to Suffolk. . . . There it is all air and water . . . here there are lots of trees and reeds. Gay men go in for aggressive nude sunbathing. They can lie in the enclosure – some of the others didn't like it so they put a wall across, which didn't suit anyone.

The men's pond has a longer history than the women's pond, attracting similar allegiance and passions from those who swim there. Those interviewed spoke lovingly of its natural delights:

> it is unlike public parks, giant graveyards, this is nature and nature looks at you, you can't look at it, it is unique in that respect . . . pagan, you could not be put in charge here, this place is in charge of you, in charge of everybody that comes here, this is where nature takes over and human nature is just watched out. It makes me ageless when I come here, but when you go to enclosed places, they have an atmosphere about them, things you cannot do, notices: don't do this, don't do that, back here it is no, it controls everything. This is where you have an awareness of being and it is people . . . vandalise it or create a disturbance, it just does not happen here, it controls everything.
>
> (Two elderly white men in their eighties)

> It is a wonderful place, this is the most wonderful place in London, unspoiled.
>
> (White male, 78, English, has swum in the ponds since 1946)

In many ways a similar sociality was enjoyed and described:

> Relaxation, truly relaxation, communication with other people.
>
> (Another respondent)

> It is unique and costs nothing and it is easy to get here and everybody is very pleasant, no quarrels, argument. It is very lovely, it is a common for the common people. Everybody seems to be on a level here, there no pecking order.
>
> (Another respondent)

> There are a lot of people that come here regularly and you start to talk to people and over the years you get to know them, so often when you come here you say hello to a few people. . . . I would say that's the advantage this place has, it is able to bring together all sorts of people, age group, gender, sexual orientations.
>
> (Middle-aged man)

> It depends on what day you come here, come on a Friday afternoon, you see more Jewish, Orthodox Jews, especially on Fridays in the summer because they finish work early and get back for Sabbath, that's an extra hour so they come here, I suppose more old people and obviously a very gay scene.
>
> (Youngish man)

> Many races in here but usually Caucasians and blacks, so we do have a mixed crowd.
>
> (Chilean, mid-thirties)

But there were also two striking ways in which normatively 'masculine' sex/gender differences were produced and performed in this space. First, there was little evidence that the space had been constructed in any way as domestic and cosy.

Second, as in the municipal Turkish baths, bodies were sexed, or, to put this another way, different sexualities – heterosexual and homosexual – were performed in this space, each explicitly othering the other: 'I like swimming outdoors, nice muscly men to look at [laughs]' (English white male, 26 years).

Another, asked why he visited the ponds, answered, 'Oh, please, you come for flesh [laughs]'.

This sexualisation of the men's ponds erupted as a highly contentious issue in the 1970s when Peter Tatchell (a London MP and gay rights activist) and others started nude sunbathing and explicit sexual practices. The battle which ensued between the old guard and the new, mainly young gay activists, found its way into the local and national press, and was finally resolved by the construction of a wall across the changing rooms which literally divided gay and straight men from each other's gaze: a wall of protection and exclusion at one and the same time (Marcuse 1995). Giles who is 63 described the changes as he saw them:

> I've used Highgate pond since I was eighteen. I now sometimes use the mixed pond, because the men's has lots of homophobia – it's provoked by older gay men. There's lots of real old queens in and out of the toilets. They used to have a screen. The changing area used to be classed as the gay area, but the straight badminton players resented men sunbathing and caused problems. There's lots of unprovoked homophobia – I have been pushed off the jetty there. Gays really got pushed into a corner. Now they fight back. If there's a problem we complain to the lifeguards or to the Corporation of London.
>
> Some men, when they hit their fifties, panic sets in and they start behaving badly. I've met yacht owners, writers, academics there. They were married, but mainly as a cover. It was the only place in the country where there was no class barrier. Then people started mucking it up by naming names, so now it's uncomfortable for closeted people.

And Brian who is 72:

> I've been going to the men's ponds for 50 years – but it's divided into nude and not nude and it's become very bitchy. I won't go again. . . . In the men's pond you used to have to be very discreet – lots of boxers and athletes used it. You used to admire their physiques, but kept your head and never used it as a pick-up area. You were very careful. Now it's been taken over by gay men, including the sunbathing area, and the atmosphere has changed.

Nevertheless, other interviews suggest that over the years a settlement has been broadly reached and though the men's pond is seen as somewhat gay, and for men operates as a pick-up place, there have been few conflicts. Men who could not brook open gay sexuality simply stopped going to the ponds, and for the rest there is an apparent accommodation of sexual difference.

CONCLUSION

These two different public bathing spaces derive from very distinct origins and have functioned over the years to produce bodies in very different ways. The statutory and municipal processes producing these sites have also been distinct. The Turkish baths have been largely situated in wider discourses and practices of cleanliness and health, while the ponds have been more strongly linked to discourses of nature, leisure and open space. Nevertheless, what emerges in the narratives recounted here are notable overlaps in the embodied and gendered performativities in these two spaces. Though both men and women performed social rituals in these spaces, there were clear differences: women tended to domesticate the space, while men tended to sexualise it. What is consistent, though, is how these bathing sites in different ways blurred the boundaries between public and private, illustrating the elasticity and flexibility of the very terms. In the men's Turkish baths, on one account, the gay sex is 'doing sex' in public and is thus affronting not because it is gay but because it is public. But from another perspective it can be seen as sex in private because these baths are hidden, enclosed and known to be places where certain agreements about behaviour are sanctioned – so bad luck if you go in by mistake not knowing the rules. What also is illustrated in these stories is the passion and pleasure associated with marginal public spaces in the city, which may not be highly visible but which arguably represent just as important a part of the city as the public spaces more commonly supported and promoted in public policy. What also emerges, notwithstanding some of the men's discomfort with blatant sexuality, is a high degree of co-existing with difference in these public spaces of the city.

POSTSCRIPT

The day before this book was due at the publisher, I revisited York Hall to take photographs. Once again on entering the baths I was overwhelmed by a cacophony of sounds, laughter and different languages, and struck by the diversity of bodies and cultural practices performed there. This time, though, there was a stronger atmosphere of celebration and anticipation. In the middle of the central room eight women sat around a plastic table covered with curries, pineapples, salads, falafels and other culinary delights. Chatting to the women I had interviewed earlier, I discovered the baths were to be closed for refurbishment as a 'spa'. Neglected for years, under new management bright new facilities were to be constructed. Articulated all around me were fears that the entrance fee would be raised and that the more formal, newly designed modern spaces would militate against their continued use by these women in the future. Would another marginal and secret space for encounters across differences be lost to the city, I wondered.

6 Invisible subjects

Encounter, desire and association amongst older people

In 1972 Simone de Beauvoir wrote:

> Society looks upon old age as a kind of shameful secret that it is unseemly to mention. . . . Sheltering behind the myth of expansion and affluence, it treats the old as outcasts. The aged do not form a body with any economic strength whatsoever and they have no possible way of enforcing their rights: and it is to the interest of the exploiting class to destroy the solidarity between workers and the unproductive old so that there is no one at all to protect them. . . . If old people show the same desires, the same feelings and the same requirements as the young, the world looks upon them with disgust: in them love and jealousy seem revolting or absurd, sexuality repulsive and violence ludicrous.
>
> (1972: 1–3)

Despite the shifts in social mores over the years since *Old Age* was written, in many respects these same words could have been written today. What has shifted, perhaps, is the force of the moral panic about the burden older people in increasing numbers are anticipated to represent, as people live for longer and the relative proportions of those employed and those retired change, and decent state-provided pensions are considered less and less viable. Thus older people are increasingly seen as a huge dependent burden on welfare and health provision and on a tax base provided by a declining work force (Katz 1996: 2).

Nevertheless, that said, there are shifts in discourses on ageing and older people, and there are signs of older people exercising greater power over their lives and over policies made on their behalf. As the vociferous and politically minded baby-boomers move towards this stage of life (Huber and Skidmore 2003), the meaning of old age will doubtlessly be recast, just as those stages preceding it have also been reimagined and lived. The notion of old age has not only changed over time, it is also culturally constructed with very different understandings of what constitutes old age in different parts of the world. The meanings of old age are thus plural, diffuse, enigmatic, contradictory and complex. But consistent themes persist: the lack of visibility, older people as undesiring and asexual subjects, old age as a space of fear and risk. Older people are represented both as embodied subjects – in the sense that they are defined by the idea of the body in decline, the body as frail – and as disembodied – in the sense that their bodies are emptied of desire and power. Katz points out (1996: Ch. 2) that the medicalisation of the aged body further 'transforms it into an inherently pathological aged subject'. In Foucauldian terms, old bodies are produced as a population that is counted, investigated, medicalised and segregated into sites characterised by poverty and need. Through these strategies gerontological knowledges and discourses of age are mobilised to further weaken and disempower an already disempowered group.

INVISIBILITY

This chapter seeks first to disrupt the myth of the dominant experience of old age being one of isolation and limited association or sociality, which, I argue, is rooted in old people's invisibility, which itself derives from two sources. The bodies of older people (and women particularly) are usually strongly differentiated from norms of beauty and attractiveness which emphasise smooth lean curves free from wrinkles, fat or blemishes. Thus they rarely feature in magazines and other forms of media which represent the pleasures and delights of sociality. Second, the notion that there are limited social relations between older people fails to recognise the shift in the site of many of their social relations into different spaces – often, but not always, spaces which are more typically defined as domestic or private. As a result, the many different forms of older people's sociality – of 'rubbing along' in city spaces – are obscured.

Two different vibrant public spaces have been selected here to illustrate my argument: the University of the Third Age and allotments. These sites offer a counter-argument to Putnam's (1993, 2000) influential account of declining levels of association. Older bodies are figured as asexual and non-desiring – representing another arena of invisibility. This notion is disrupted by the narratives of one group of older people, a group even further marginalised from view – gay people. The final part of this chapter revisits the theme of risk and fear which runs through many of the chapters of this book.

THE UNIVERSITY OF THE THIRD AGE (U3A)

The University of the Third Age represents one of the fastest-growing social movements in the UK over the last quarter century. Yet it is largely invisible except to those involved. The term 'third age' was first introduced to Britain by Peter Laslett in 1981 when the first British University of the Third Age was founded in Cambridge (Laslett 1996: 3). Laslett argued forcefully that outmoded attitudes, stereotypes and assumptions continued to distort the image of older people, one reason being the rise of scientific medicine of the early twentieth century. Though this was not explicitly articulated, the use of the term 'third age' was clearly a discursive strategy to disrupt normative attitudes to older people and ageing and create a new collective identity. The second age is defined as a period when people have full-time employment and family responsibilities, when work is wholly imposed by others and allied with loss of time. This leads, according to Laslett (ibid: 188), to the notion that all people want to do in retirement is take a long rest, the dominant idea in the 1950s being that this was the time to pursue gentle hobbies. With growing longevity, the third age for many people may extend for an equivalent period to the second age, as the (traditionally masculine) boundary between public and private time – employment and retirement – is gradually breaking down. The 'third age' is deployed to replace terms like 'old age', itself a homogenising category, to stand for 'dignity and creativity, the social and public significance, the self respect and civic virtue of older people which certainly continue indefinitely into later life' (Laslett 1995: 10). This is argued to be a time of personal achievement and fulfilment which is not defined by chronological age.

Another proponent of the third age, Eric Midwinter (1992) proposed the idea of the outlawed citizen to describe the exclusion of older people from civic and voluntary life. As part of the 1990 Carnegie Enquiry he reported on the many and various forms of exclusion: the Citizens Advice Bureaux until recently took no one over 65 and imposed obligatory retirement at 70; the Lord Chancellor's department stipulated candidates for the magistracy must be below 60 with preference for those under 50; jury service is curtailed at 70; other enforced retirements included: Guides and Scouts – 65; British Red Cross volunteers – 75; St John's Ambulance – no new volunteers over 65; lay magistrates on juvenile panels – 65; industrial tribunals – chairmen 72, members 68. Thus, Midwinter established that people over 55 were less involved with volunteering by a ratio of 1:4 than the rest of population. Underpinning these exclusions, he argued, lay the assumption that chronological age can be used as an indication of capacity, denying the obvious diversity in this population. In the place of chronological age, Young and Schuller (1991: 22) argued for new images or metaphors of the life course, suggesting that the medieval use of the circle to depict the life course could replace the twentieth-century image of a line.

The University of the Third Age was initially based on the French Université de Tous les Âges, established in 1972. It was founded in the UK in 1981 by Michael Young, Peter Laslett and Eric Midwinter, 'who had the temerity to challenge the British educational and political establishment with the disruptive notion that retired people had the wit, experience and energy to manage their own social and educational affairs' (Midwinter 2004: back page). The U3A had a number of objectives (see www.u3a.org.uk), summarised briefly here as:

1 to educate British society at large about its age constitution and the implications of this;
2 to enhance older people's awareness of their intellectual, cultural and aesthetic potentialities and to assail the dogma of intellectual decline with age;
3 to draw on the resources of older people in an affordable way for the development and intensification of their intellectual, cultural and aesthetic lives;

4 to create an institution for these purposes where there is no distinction between those who teach and those who learn and where all labour is freely and mutually offered;
5 to create an institution where learning is pursued with no reference to qualifications and interests are developed for themselves alone;
6 to mobilise university members to broaden the educational opportunities of other older people who want to engage in education;
7 to undertake research on the process of ageing in society;
8 to encourage the establishment of U3As across the country where conditions permit, and to collaborate with them.

Thus the University of the Third Age – unlike the Francophone model from which it drew its inspiration and which more closely resembles an extension of university – embodies a culture of mutual aid and voluntaryism in adult education where those who teach also learn and those who learn also teach. This is a crucial element of the organisation, and all the people interviewed commented on the enhancement of members' (particularly women's) confidence through this process. At the same time U3As make 'older people more visible, creating a public space for generations whose presence can be overlooked and whose worth is often underestimated' (Laslett 1996: 31). By 1995, 1,500 U3As had been set up in over 20 countries, including 290 in France, 400 in China, 17 in Poland and 250 in the UK. By 2004 this had expanded to 552 U3As in the UK involving 1,421,777 people (U3A website). U3As are set up in local communities by local residents, often by word of mouth. Classes given depend on local interests and expertise and can range from architecture and history of art, to computing, cooking and poetry reading. Other classes are aimed at creating and maintaining healthy and flexible bodies (Sources 2004). Annual fees are kept to a minimum (e.g. £20 in Highgate).

In Britain U3As form the intellectual vanguard of the third age as a whole. The organisation is also concerned to make a political intervention. As Groombridge (1995: 29) put it:

> It has become more and more obvious in our time that government has become impossibly difficult. That is one reason why there have to be Non-Governmental Organisations representing thoughtful, well informed citizens at major events such as the Rio Summit. The responsibility for policy-making has to be more widely shared. Democracy can no longer be equated with choosing a handful of people, most of them middle aged men, most of them no brighter than we are, to decide everything. That is the reason why U3As must get involved in a more deliberate overt way.

Party politics within groups are seen as less desirable. Talking of her South London group, one respondent, Ruth, said:

> Politics we steer clear of. . . . We do try very hard to keep off politics. Two members of a far right organisation caused a lot of upset. In the end the

> Chairman said, 'If you carry on like this the group will fold.' The group didn't fold. They did carry on, but after a while they left. Great relief.

And Sal from Rhondda U3A (34 members in 2003): 'We keep politics out of it altogether. Lots of politics, you get dissension. Probably in our group most people are Labour. I'm not and they know that.'

For many the U3A is centrally a space of sociality – 29 per cent in a 2001 survey joined to make social contact and 53 per cent cited companionship and meeting others as the aspect of the organisation most enjoyed.

In the words of Maud (South London): 'Loneliness is terrible. U3A combats that. It stops people being lonely and they live longer. . . . It is a godsend. An absolute godsend. I don't know what I would have done all these years.'

While in Edith's view: 'For many women who lose their partners it is a lifeline.'

And Bill from Richmond U3A (with 291 women, 82 men in 2003): 'We have a very positive influence in "maintaining people's marbles", in "social therapy" and in keeping people connected.'

U3A represents a temporal and spatial reordering of public/private space, challenging the boundary between them. The majority of classes take place in members' homes, reconstituting domestic space as public for the period of the meeting. It is this domesticity and privacy which construct and reinforce the lack of visibility of U3A to those not involved. Many of the participants – 73 per cent (TAT-NEC 2001) – first heard about U3A by word of mouth, a process which inevitably produces sameness and homogeneity in members to a considerable extent. Class and race are crucial here. Members of the U3A in the UK are almost exclusively white (in the survey 96 per cent) and often, but by no means entirely, middle class, with a majority of women (74 per cent). In answer to questions about the social composition of U3A, ill-informed and apparently unconscious othering of those who are not white was frequent. Maud again:

> There are enough men in the U3A for men not to feel uncomfortable about belonging. There are all sorts of classes and a tiny sprinkling of coloured people – a matter of regret nationally and locally that we don't seem to attract them, whether they have . . . [trails off] Maybe they don't speak, or they don't have the confidence or they're just not interested.

In Ruth's talk, the lack of participation of Afro-Caribbeans appeared as a fault of 'them, not us':

> I don't think U3A is what they want and we can't be what they want unless one of them comes along and offers to start a group. We'd like it to be more mixed. We are slightly mixed. We've got a number of people who are white but foreign. Coloured, well I don't know because I don't see all the members. Some. Mostly Indian. Can't get anywhere with Afro-Caribbean.

Sarah's comments from Kingston U3A (647 members in 2003) illustrate how acceding to British dress norms facilitated one woman's acceptance:

> Race, that's a ticklish one. A couple of years ago we thought of going round the mosques. One or two people, not African or Caribbean. Indian extraction. Asian. A lady from Thailand joined. I spoke to her on a walk. Nobody has accepted her. I thought that was a bit sad. An Iraqi woman. We're all very fond of her. And she gets on with lots of people. She's not dressed in a burqa kind of thing.

The only challenge (at the time of the research) to this hegemonic white positioning of U3A in Britain had come in 2000 from a group in Tottenham, North London, initiated by a politically active Asian woman in the local council who called a meeting in the local Marcus Garvey Library encouraging local black contacts to attend. The initiative was immediately successful. At the meeting 25 people were present – five men, twenty women – of whom two were Asian, seventeen were Afro-Caribbean or African, and six were white. The Chair of the Committee, who had come to the UK from Jamaica in the 1950s, made some interesting points:

> we cannot have the meetings in our houses. Often they are too small, but also for us hospitality is very important. You can't invite people to your home and not give them tea and cakes and other things to eat. The same goes for our Asian members. So if you had the classes at home you'd spend the whole time in the kitchen, or preparing food in advance, and never get to the classes.

For this group, then, success was dependent on the offer of free, local and accessible municipal facilities where classes could take place. The contrast with another group in a wealthier part of the borough, Highgate, highlighted the hidden racialisation of this avowedly democratic culture. The Highgate U3A is exclusively white and mainly middle class. The majority of meetings take place in members' houses, although a number of popular courses are run in a local National Trust property where a small fee is required. The architecture course arranges overseas trips where clearly access to finance represents a prerequisite.

The fact that the space of U3A is private and domestic legitimates and reinforces further exclusions. As Dorothy (South London) put it:

> We can't accommodate people in wheel chairs since many private houses don't have disabled access. Once someone wrote a letter to ask to do Spanish – said he was a recovering alcoholic. We really can't cope with recovering alcoholics or drug addicts. Well, alcoholics – that's fine. It was the recovering drug addict. We have even had people who are referred by psychiatrists.

Despite limited recognition of these issues, there have always been members of U3A advocating diversity. At a national conference in 1989 Futerman (1989: 10) argued that:

> U3As and similar organisations are springing up throughout the world and part of the strength of that community lies in its diversity . . . like all strong

> communities the differences between the members are not only tolerated but welcomed. . . . It is very much hoped by the Organising Committee that this sense of belonging to a global family will be enhanced by the opportunity to gather together at many more such Symposia.

The U3A, then, is an association whose collective identity is formed through a number of discursive strategies deployed in mainly local state and community arenas. The members of U3A have, in a sense, constructed an alternative space to the dominant discourses on age and created a new public space of sociality for older people. Since its formation, a group of people – particularly women – have come to redefine themselves as 'third agers' and have gained confidence, new skills and knowledge as a result. More recently initiatives to build a virtual U3A represent a further move towards inclusion of those spatially marginalised people at the same time as facilitating older people's participation in the Internet – a sphere from which they are often excluded. Its composition is fluid and growing, its culture, democratic and participatory. For many, U3A represents a vital and positive public space where few alternatives exist. At the same time, though, an entrenched sense of whiteness and white culture as the norm, 'othering talk' by some of the respondents interviewed, and a spatial form of sociality relying on the availability and accessibility of private space construct and reinforce the exclusion of non-white participants.

ALLOTMENTS

Allotments are invisible in a different way. They are not domestic. Instead they occupy marginal spaces, they back onto the railway track or the canal, are often not visible from the road or are at the back of houses with only a gate onto a street alerting the passerby to their presence. Allotments have a long history. In 1819 an Act of Parliament was passed linking allotments with the Poor Law and empowering church wardens to buy or hire land to let out to the poor and unemployed (Butcher 1918: 11). As a result of the involvement of the Labourers Union and three further Acts in 1887, 1890 and 1894, parish councils were empowered to hire out land by voluntary agreement, culminating finally in the 1907 Small Holdings and Allotment Act. By the start of the First World War there were 600,000 allotments, increasing to 1.5 million by 1918 (Bullock and Gould 1988: 9) as they played an important role in providing much needed fruit and vegetables for the nation. Between the wars the number of allotments declined, despite their importance to many families during the depression as a food and recreational source, to pick up again during the Second World War. The post-war period witnessed another decline owing to the availability of cheaper vegetables and high levels of demand on land for house building. However, interest in their provision surged during the 1970s, waiting lists for allotments doubling between 1973 and 1974 to 57,000 (Friends of the Earth 1974: 3). By 1988 half a million households in the UK had allotments (Crouch and Ward 1988: 16). In the latter part of the twentieth century allotments were drawn into various movements for urban open space and land rights campaigns like 'The land is ours', while studies of the regeneration of allotments highlighted their significance as social and environmental resources (Wiltshire *et al.* 2000; Wiltshire and Crouch 2001). Groups acting to promote Local Agenda 21 – an initiative from the 1992 Earth Summit – like the Quality Environment for Dartford (established in 1996), have been active in promoting and supporting allotments, especially their role in sustainable development strategies. Emphasis on allotments as sources of food has also been highlighted (Howe 2000). Despite this support, and growing waiting lists in many large cities, allotments remain under threat as councils and private developers make applications for construction on the land – 50 such applications were made by local councils in 2004 (*Daily Telegraph*, 9 August 2005, p. 7).

Though originally allotments were a predominantly working-class pursuit, over the years they became popular more widely and were articulated by many as a space of democracy, self-help and co-operation. Butcher (1918: 32) describes the prevalence of 'doctors, parsons, shop keepers, policemen, postmen, engineers, electricians, journalists, civil servants, clerks, people of independent means, labourers of every class'. That may be. Nevertheless the allotment has persistently remained a highly gendered space, 3.2 per cent of allotment holders in 1969 being women of whom 1.85 per cent were housewives (Crouch and Ward 1988: 91). Butcher certainly saw women's presence as worthy of comment: 'there are instances on record, too, of women who have never previously enjoyed manual labour of any kind, who have yet succeeded splendidly' (1918: 35). One such woman's

> efforts were received by her fellow plot holders with some amusement, and a good deal of disapproval; but, not dismayed, and rejecting with scorn repeated offers of help from the men, this Amazon of the soil persisted diligently and methodically in the heavy task of trenching.
>
> (ibid: 36)

Over the years older, predominantly white men remain the strongest image of the allotment holder, and this group still probably constitutes the majority, though figures are hard to find. Allotments thus represent an important and also relatively invisible public space for older people, and older men specifically. Several allotment holders were interviewed for this research in London. Tessa is 65. She has lived in London all her life. She has never been abroad:

> I've had an allotment on this site for seven years and this is my second one. I got the first one when I retired. I saw an advert – I think in *Haringey People* magazine. I came up and met George, the secretary (a Cypriot), and went on the waiting list. It didn't take long to get, a few months only. It took longer to change to this one, about a year. I needed to change because my neighbour on the first one kept interfering. I don't know if it was because I was a woman, but he was always trying to show me how to do things. I think he liked me, but it was overpowering.

I haven't got a proper garden where I live – just a pocket handkerchief. That's why I got the allotment. I live on Noel Park Estate (just off Wood Green High Road). I hate it. There's a nightclub nearby. I hardly get any sleep at weekends because of all the people coming out of it at about 4 a.m. I used to have a small garden. I grew a few flowers, but mainly used it to sit and drink tea. It was my little private space. I wasn't a great gardener – I'm not one now. . . .

A garden is convenient because it's on top of you. To get to my allotment I have to take the bus, but as soon as I'm through the gates I'm in another world – it's such a peaceful atmosphere, there's so much space, loads of room to grow stuff, nice neighbours and people to talk to. I get fresh plums, apples and pears with no pesticides – though the birds get a lot. I love to sit and watch the birds and squirrels – it's nice to see them. . . . It's an escape from the rat race. It stops me getting old and just sitting in front of the TV. There's also company – I don't get that at home. At home I think of my age, probably because I'm bored. I also think more about my aches and pains. I come here to feed the birds in winter – usually I walk for the exercise. There's a robin nesting here. It brought up chicks – beautiful to see. It landed on my head once.

The allotment keeps me fit. I'm tired when I go home. It keeps me active – sometimes I'm ready to drop. I rebuilt my shed and got a gas burner, so that I can make tea.

. . . I've learned everything since I've been here – I knew nothing except flowers before that. People did volunteer to help me – probably because I'm a woman, they thought poor little woman and helped out. I didn't mind that. You can learn from seed packets and books, but people with lots of experience are better to learn from.

I think it goes back years, when allotments were for little elderly men. We've gone past that now. Women are not so helpless as they used to be. I'm not a little weakling. Being a woman doesn't affect how I socialise here. Ben, my neighbour, talks to me about anything. He knows I'm not a moron. I help him with tasks and he helps me. In fact I'm more agile than him.

There are all nations here. I think Great Britain is too overcrowded, but at the allotment you meet people more individually. You see a person rather than having an unknown shoved on your doorstep, so you get to know them. There's good and bad in all nationalities and races. I think I mix OK. There's quite a lot of young people here over the last few years. Seven years ago it was mainly elderly men. It was rare to see a child – now there are lots more children.

I don't think kids should be at allotments. They are somewhere for grownups to relax. There are lots of children's parks – I wouldn't go to them. This is our pleasure – grownups are entitled to their own space. We did set aside the front allotment for people with learning disabilities. They came a bit last year, but haven't come this year. I suppose that hasn't worked.

Plate 6.1 Women growing herbs and spices that are difficult to find locally on a London allotment.

> I hate people stealing from allotments. Most people here are poor. It's unbelievable that people can be so nasty. They should be slung off the allotments.

Tom is 79 and has lived all his life in London. He was a foreman in a timber yard till he retired:

> I've had an allotment here for 40 years – it cost £1.49 a year then. There are nice people here and I'm happy to give my shed keys to other people. I don't have a garden – I live in a flat. In fact I've never had a garden. I got an allotment because I wanted somewhere to occupy myself. I like it here. There's a lot of camaraderie. Indoors – at home – it's just the family, now just my wife and me. I wanted to get away from the hustle and bustle of home and have somewhere to relax.
>
> I like it here because it's somewhere to go. It's a place to relax with lots of camaraderie and helpful, good neighbours. I do it by myself. My wife used to come over, but we couldn't agree what to do, so she gave up. My grand-daughter came once or twice, but she got bored. Kids do.
>
> Social life? Definitely. I sometimes share drinks with people. We sit down and chinwag. If someone isn't well, I'll help out with digging, seed sowing, watering, etc. I know that people would do the same for me.
>
> I've learnt practically everything here and a lot from other people. I had no experience of gardening before getting the allotment. I find there's a great atmosphere here. It's an outlet for me. I don't go to the pub – I only drink lemonade, so it's a good place to relax, meet and chat with mates. The camaraderie is great – the next door person dug my allotment when I had a bad leg.
>
> I can't think of anything bad. There's a good secretary and a good committee. I've never had a robbery from my shed – in fact I share it and let other people have a key. I've never had people nicking produce, except once, years ago.
>
> I think it's mainly men because women prefer to be at home. I personally don't care whether it's women or men on the allotment.
>
> There are more people from different cultures and backgrounds than there used to be. I don't think it has made much difference. I try and get on with everybody – live and let live.

Sarah is 68. Until she retired she was a social worker in Camden. She has lived in North London all her life and in her current home for 34 years. She lives alone. She has a partner who lives in Ealing and they spend weekends together:

> I love my allotment, which I share with my partner in Ealing. I spend hours and hours there. I like growing things, then eating them. Our allotments don't allow you to use hosepipes – I think that discriminates against old or frail people, having to lug buckets back and forth. There's a minor social life there.

Plate 6.2 An allotment at the Kentish Town City Farm, London.

> People give each other plants and share produce with one another. Our neighbour was ill for some time – has only just come back. We watered his plants while he was away – but also ate his beans. People there are universally friendly and we do learn from each other. It's mainly men, but that is not an issue and I don't think anyone knows or cares whether we are lesbians. It's not at all culturally diverse – very Irish and UK. That's maybe because it's not run by the council – I think it belongs to a church.

Tessa's, Sarah's and Tom's talk illustrates the gendered nature of allotments – both in terms of the way women are specially treated as allotment holders, and in their constitution as predominantly men's space.

Martin is 72. He has lived in London since 1968. He used to be a teacher until he retired in 1996. He too refers to the significance of gender in his use of allotments while also drawing attention to issues of cultural differences:

> I've had an allotment since 1970. To begin with, the allotment was a balm to heal the stresses of teaching. It was a contrast from my day-to-day way of life. Teaching was a hurly-burly existence. I worked hard . . . it became stressful and I used to get very wound up. The allotment let me wind down and meant that I didn't take all my stress home. . . . I think the allotment is therapeutic in repairing the damage caused by the stresses of everyday life. It satisfies an instinctive need to be in touch with nature and with the process of sowing, growing and harvesting. . . . My plot is a place to go when I need to get away from the house. It allows my wife to have more space for herself and lets me create space for myself. It's an opportunity to mix with people from different countries and many different social and cultural backgrounds. I didn't get an allotment with that intention – it was incidental, but it's an important by-product.
>
> Yes I have a social life on the allotment. I've built a circle of acquaintances and friends. It doesn't spill over outside the allotments – I don't socialise with anyone here outside, but I really enjoy chatting to people here . . . I've become aware of different ways of looking at things – partly to do with plants, but partly how to look at the world. . . . The allotments have become more diverse over the years. On the whole I like that. I don't find it threatening. Some attitudes and practices within some ethnic groups can be irritating – like the lack of participation in running the allotments, or the reluctance of some to follow a few rules for the benefit of all. It would be nice if it was a melting pot here, but it's not . . .
>
> One of the problems is that different groups won't integrate. They keep themselves to themselves and barriers go up. There aren't enough people willing to participate in running the allotments and I don't know what the answer is. Most allotment holders are men. I think that's partly down to social structures. Men have time, especially in present society. Nowadays men and women both work, but women also have to look after the kids, do the housework and cook the meals. A man usually does less of these things and

> so has more time. That pattern is taken forward when people get older. . . . Mainly I like the diversity and contrast of the allotment with every part of life. It really is a special place.

These two sites of sociality, sites of enchanted urban encounter for many, disturb, at the very least, and contradict, the vision of declining social capital and association proposed by Robert Putnam (1993, 2000) which has been so influential in the USA and Europe. In *Making Democracy Work*, through a study of the twenty regional governments in Italy in 1970, Putnam establishes a connection between strong networks of civic engagement such as neighbourhood associations, co-operatives and sports clubs characterised by intense horizontal interaction, and formal democratic participation. For Putnam these horizontal networks are an essential form of social capital which fosters robust forms of reciprocity, social trust and norms of acceptable behaviour which are mutually expected, performed and reinforced (1993: 172). The more communities participate actively in mutual exchange, displaying their trust of one another, the more other forms of social capital will develop. Thus in social contexts where group members trust one another, sharing their skills, labour, knowledge and tools, a strong society, economy and state will ensue. In the Italian context, Putnam found the regions with a large number of active community organisations, such as Emilia-Romagna and Tuscany, were also those where democracy appeared to be working. In contrast, 'uncivic' regions like Calabria and Sicily were revealed to have limited engagement in social and cultural association and less effective representative government.

In a more recent text, Putnam's (2000) concern is to explain what he perceives to be the decline in civic engagement and erosion of social capital in the USA over the last three or so decades. Drawing on the metaphor of 'bowling alone' (the book's title), which literally here describes the kinship felt between two members (one white, one African-American) of a bowling club where one man donated a kidney to the other, the book more broadly argues that Americans need to reconnect with one another and rebuild social capital. That these two men 'bowled together made all the difference' says Putnam (ibid: 28). From a detailed study of Americans' participation in a wide range of arenas – from public politics to clubs and community associations, religious bodies and work-related organisations – Putnam draws the conclusion that following the deep levels of engagement in community life in the first two decades of the twentieth century, in the last three decades people were pulled apart from one another and from their communities. What is missing in Putnam's argument is an exploration of association which occurs within different or less obvious spatial forms, whether domestic, interstitial, temporary or fluid. These are precisely the public spaces often inhabited by those more socially marginalised like older people. So Putnam is looking in the wrong places.

UNSEXY BODIES

Simone de Beauvoir put it baldly: older people's sexuality is seen as repulsive. Though research suggests that older people's sexuality and sexual practices are as heterogeneous as those of people at any stage of life (Walker 1997), and initiatives to empower older people to articulate their sexual experiences and desires are now more commonplace (Blank 2000), older people's sexuality remains barely visible in most societies. Where people get together in later life, it is described as either sweet or absurd and also as disgusting. Perhaps the very discourses surrounding older people's sexuality produce the effect of invisibility, the need to hide away, in those desiring subjects, which in turn makes its public occurrence and visibility all the more strange. So public space for older people is not a space of sexual expression, and public spaces for sexual encounter barely exist. Nowhere is this more true than for gay and lesbian older people. Despite recent shifts in the sphere of representation and daily urban life in some cities, where gay identities are performed in certain localities like Greenwich Village in New York (Chauncey 1994), Castro in San Francisco, Darlinghurst in Sydney and Soho in London (Aldrich 2004; Mort 2000), there are still many parts of the world where gay identities are hidden and excluded, if not literally harmed and violated. And despite the political campaigns of groups like Polari in the UK, older gays are largely written out of the public realm. To be older and gay is to be doubly invisible. For this reason I have chosen here to draw attention to older gay people in urban public space. Questions of visibility, privacy and publicity ran through many of the interviews.

Howard, 58, works in the supplies department of a London health authority.

> If I'm on the bus with a friend who is also gay, I wouldn't say anything to show that we were in case people heard. I am not ashamed, but I don't know what people's reactions would be. At home I wouldn't worry. I'm in the LGBT group of my union and after meetings we sometimes go to a gay pub. I went to gay bars more in the seventies. It was a novelty then. Now I don't want to go out every night. I suppose it was an obsession then – coming out, finding myself, bars helped . . . I don't like today's gay clubs. They are expensive, loud and noisy. Some gay bars are so expensive you'd need to take out a mortgage to go to them.

Gill is in her late fifties. She moved to London to be with her partner and is a self-employed support worker for Hackney Parks Forum, which maintains Hackney parks:

> As a lesbian, private space gives me the freedom to be who I am. When friends visit, I don't have to worry about my sexuality. There's a level of relaxedness and openness.
>
> In public spaces, most people – not just lesbian and gay – present a persona which isn't the same as when they're in private. With my partner in public,

> I think I'm obviously lesbian even though I'm not stereotypical – though probably that's not apparent to other people except lesbians. I think this public persona thing could be a white English cultural issue. We have a façade of public image. Other cultures are likely to be more overt and display their true image. White English people are much more reserved. In other cultures lesbians and gays may have more of a struggle because the subject is more taboo. Gay Pride brings out the best and worst of everyone. It's the one place I feel free of those shackles. I went to Sydney Mardi Gras five years ago. Seeing the police marching and watching the spectators applauding was really liberating.
>
> In private there can be complete relaxation. I share intimacy with my partner, we use a different language, have pet names, I say darling all the time. In public my partner would get very embarrassed by that. Intimacy is held back in public.

Sarah is in her fifties and has lived in London all her life. She lives alone and she develops training programmes.

> From a lesbian and gay aspect – intimacy, like holding hands, tends to be in private, except in Old Compton Street. Heterosexuals can do that in public. At home, with friends or a selected audience, I can do anything. I've had a lesbian Anne Summers party – there's no other space we could express ourselves that way. Sex has to be conducted privately, especially gay sex – in private anything goes. There's a whole other world that people don't know about – AA, the lesbian and gay scene, coming out issues.
>
> . . . I've been in a lesbian group for eight years. We started by meeting in people's homes. Now we go out a lot. We have curry nights, where we take over the back of a restaurant. It's obvious that we're dykes – we're open about how we express ourselves, the same as if we're at someone's house – but it's all very easy going. We go to a pub every Thursday to play snooker. Not gay pubs – because it's a student area everyone's laid back. We mix teams with the young students and all get on. We go to the theatre, cinema – especially the Lesbian and Gay Film Festival – and play tennis.

Graham is 63. He has lived in Hackney all his life. He now lives in his grandparents' old house. He used to work in advertising, until he retired following an accident two years ago. He now writes articles and fairy stories. He has been with his partner for nineteen years, although they do not live together. He is out to his family:

> In public I worry about being open, e.g. even kissing a man on the cheek, though it's OK to kiss a woman. If society was as we'd like it to be – if hets can express themselves in public, then so should gay people. They wouldn't need special places.

Adam is 72. He grew up in Portsmouth and has lived in London since the mid-1950s, in Hackney for the past 30 years. He has been out of work since he was 50. Before that he worked for banks and insurance companies as a messenger and claims clerk. He had a heart attack a few years ago and since then has lived in sheltered housing in Hackney:

> I always think of Speakers' Corner as a public space, which I always associate with gay cruising. Other places for older gay men to meet are PACE, a counselling service, South London Gays, Pimpernel, where there's a group of 25 over 60s who meet every Thursday, Marypad, an ex-CHE group and the Monday Group, another ex-CHE group which meets at Central Station, Kings Cross. Also Gay's the Word.
>
> In public places I'm very wary. There's always a feeling of hostility or potential hostility. I'm not at ease with people – I watch how I look, walk and speak. This could be anywhere – on the street or in the supermarket. In any situation I have to decide whether or not to come out. I used to live up the road when I first came to Hackney. There were always kids around. I had to be very careful, because with kids around you just don't know. It came to a point where I contacted Hackney about sheltered housing. The warden is happy – she has a gay man living with her. I'm not necessarily open with the other tenants – only if they ask. It's difficult in sheltered accommodation because the women here are always trying to pair you up. One of my aims eventually is to encourage more gay people to move into sheltered accommodation – that's why I'm involved with Polari. I'm always careful in public. When I'm out with people who know I'm gay I seem to come alive – I'm in a different world. I don't really use public places. I used to go to Speakers' Corner, where there was someone who spoke about gay issues. I used to go and take photos. The speaker stopped going, so I don't go now. . . . Sometimes I smile and nod to people, but not all the time because I might do it to the wrong person. I don't necessarily look gay but I'm afraid that if I drop my guard I'll be identified. I feel like I have to wear a mask all the time.

Daisy is 68. Until she retired she was a social worker in Camden. She has lived in North London all her life and in her current home for 34 years. She lives alone. She has a partner who lives in Ealing and they spend weekends together:

> Being a lesbian, I am inhibited about public displays of affection, while I'm not in private. Sometimes I go to church – specifically to be with people and take part in a communal act of worship. I go into town, St James, Piccadilly, which has a thriving lesbian and gay congregation. I meet friends there and we socialise after church – often go for a meal afterwards. I don't go to restaurants much. My partner worries that people will look at us and say, 'There are two lesbians'. I think they'd look and just see two elderly women. I don't go to clubs because I can't stand the noise. I go to pubs – I especially like going out to country pubs. I don't go to lesbian pubs. They have

> terrible beer. I did go to one last Saturday. There was a disco called Sappho revived. It was mainly old Sappho aficionados – it was fun.

Kevin is 62. He has lived in London for 40 years, initially in bedsits in various boroughs, from 1977 to 2001 in Tooting. He moved to New Malden when he retired in 2001. He was a social worker in Camden for 35 years. He has lived with his partner since 1970:

> I'm fairly out, though slightly cautious about what I say and where I say it. Some people at the court (where I work as a magistrate) know, though they don't ask questions – I choose who I come out to. I wasn't out when I started in social work. Some people had problems with it – there was the whole issue of paedophilia – but then a friend helped me come out. I found gay people in the office, then it stopped being an issue. I need to feel quite secure to come out – once we met people on holiday and got on well. When we told them, they didn't want to know us any more. I don't flaunt it, impose it or behave outrageously in public – but I might if I was with gay friends. I'd feel freer.

Marje is 78. She came to London in 1946. She worked as an editor in an academic publishing house till she retired aged 65: 'I think of lesbian venues as being more like a private group space – I suppose most public spaces exclude some people.'

Dean is 72 and lives in a Soho flat. He is an ex-actor.

> I go to a café in Old Compton Street called Dukes. It's nice – doesn't have Gay written all over it and you're not surrounded by evil queens. I like gay bars with a good cabaret like the Quebec and places which aren't too crowded. Soho is like a chameleon, it alters personality. It's bright and cheerful by day, sinister in the evenings. I'm happier being an older gay man than when I was younger, because the world has improved. Even so I'm more privacy conscious – it's nobody else's business . . . I think you always feel slightly on the perimeters of society when you're gay.

The responses here are ambivalent, with varying degrees of ease and caution expressed about being gay or performing a gay identity as an older person in public. Though gay people have become more visible and accepted in some cities, and commercial interests have proliferated to capitalise on the 'pink dollar', with gay bars, clubs and cafés where gay sociality is easy to find, gay spaces in central cities are predominantly figured as youth space frequented by an often wealthier crowd of people working in the new financial services, well-paid professions and media industries. They are not typically places where older gay people feel at ease. Here is Sarah again:

> I used to go to clubs and bars, but not now. Clubs are on the young side and smoky venues are not comfortable – that's OK, it's my choice. I go to pubs for quiz nights. It's fun and something different to do.

Dean again:

> I very seldom go to gay pubs. I don't agree with them because they're overpriced and fill the pockets of straight breweries who exploit gay people. London Gay Pride also exploits gay people – Brighton Pride doesn't, it's free. . . .
>
> Older gay people are not catered for by central and local government – there are no luncheon clubs, Hackney Older People's Forum is dreadful – they've never mentioned lesbian or gay issues. Hackney doesn't do enough for lesbians and gays – they talk diversity, but don't do anything. I don't think gay people are part of the larger community.

Brendon is 72 and from Glasgow. He has lived in London since 1955 and was an accounts clerk with the local council until retiring at 65: 'Gay drinking clubs used to be fun. They've changed. I couldn't go to Heaven – all the heat and bodies. I like the Quebec – because everyone's old and disintegrating. I feel at home because I'm in that category.'

Kevin again:

> I don't always feel comfortable in gay bars – I feel ancient. When I was in my twenties I was aware of hostility to older gays – I had a similar perception, but I liked older people, so I wouldn't be as derogatory as other people. Nowadays there are places which aren't targeted at young people – like the City of Quebec and King's Arms. They have a much wider range of ages, which makes me feel more comfortable.

It is clear from these comments that many gay older people inhabit the margins of an already marginal group. Though gay people are increasingly visible, exercising power, through money, in urban space, older gay people remain largely invisible, their spaces of association hard to find. In common with allotments and the U3A, the gay sites of sexual contact, sociality and organisation are often only accessible to 'those in the know', with the concomitant internal social homogeneities excluding those who are different.

A STATE OF FEAR

Virtually all the older people referred to issues of safety, risk and fear. Beck's (1992) notion of the risk society is relevant here, where social agents are increasingly forced to make decisions for themselves on a highly individualised basis. Though in Beck's account physical harm more often derives from technological and industrial developments, a more pervasive sense of fear of harm arguably constructs everyday life, particularly for the more vulnerable. Giddens (1994, 1998), as we saw in Chapter 4, similarly points to the new moral uncertainties, and our need to identify, confront and control perceived risks, a set of actions which have become routine (Mythen 2004). The extent of increased dangers in the everyday lives of older people is impossible to assess, but the proliferation of risk discourse and sensationalised media reporting of large and small dangerous events, assaults and accidents have created a climate where older people typically experience a sense of vulnerability in public space. For older people this fear is often located in the figure of the youth who is seen as a threatening presence in public space. Bodily practices of taking up space, shoving or pushing, the performative practices of body piercing and tattooing and certain styles of clothing are seen as alienating, and thus frightening, to older people who cannot easily read or make sense of these codes.

Public transport represents another site of fear. Doris, 69, an ex-administrative manager for Haringey Council:

> I feel unsafe on the underground when it stops in the tunnel. When I'm driving and come up to a traffic jam I feel vulnerable and unsafe. I don't know if I'll be able to get out. I'm also scared of road rage – I was in an incident with a lorry on the North Circular, where I thought the driver was going to ram me.

Phyllis who is 78 and lives in sheltered housing expressed similar fears: 'I feel unsafe where there are crowds or rowdiness and in places with poor lighting. Sometimes it's unsafe on buses, when I'm standing and people are crowding me. I'm afraid I'll be knocked down.'

Plate 6.3 Threatening space: a walkway in Milton Keynes.

Better Government for Older People, funded by the Department for Work and Pensions and by subscriptions from local authorities, was established in 2000 as a partnership between health authorities, local voluntary organisations, local government and older people to improve quality of life. Concerns around public space, safety, transport, spatial mobility, street lighting, dark underpasses and the threat of violence are all articulated by BGOP members as central concerns. The perception of danger – real or imagined – can trap older people into the private sphere, reinforcing their isolation and the sense of public space as a space belonging to others – and most particularly youth.

In Bob's words: 'The fear of being mugged builds up in a person . . . it becomes more and more real . . . even if it's just a perception.'

And Jean:

> I go to the cemetery on a regular basis – the other day it got terrible . . . there was a man in the distance gradually making his way towards me . . . and I pretended to look at some grave stones . . . when he got really close I rushed to my car and jumped in and locked it . . . I haven't been to the cemetery alone since . . . I wait for my family to go. I feel very guilty. I don't know if it was innocent or not. But I heard people go there to take handbags. I saw it in the local newspaper.

Thus media discourses of fear produce fear in older subjects, restricting their sense of mobility in public. Fear amongst older gays is further exacerbated on the one hand by a fear of hostility from younger people in gay public spaces and, on the other, by the threat of homophobic violence directed towards them.

Gill again:

> I don't go to lesbian and gay pubs or clubs any more. I don't feel safe there. Maybe I'm subconsciously threatened by the environment. I've not had good experiences in London – it's felt like a meat market, that you are being assessed. Maybe it's also about my age. Women of my age are not expected to go to these places and are not accepted. In the 1970s and 1980s you could be in a lesbian and gay club and feel like it was your own space and completely safe.

And Giles:

> There's only one gay pub in Hackney – it's not safe at leaving time. I know two people who've been mugged. I feel unsafe in Hackney pubs – why do people think it's OK to mug gay people and not other groups? I've been mugged four times in the past twenty years, at Seven Sisters, Turnpike Lane and twice at Stamford Hill in the early evening. The streets round here aren't safe, especially when there are groups of black people. Everyone I know has been mugged, or knows someone who has. I feel unsafe upstairs on buses. I've seen people being fleeced twice. I also heard a black conductor shouting

'batty boy' at a middle-aged man. I feel open hostility from Afro-Caribbeans. I've been spat at and punched coming out of gay places. Once I had to go back in and call the police. I do feel safe in Central London. I also think you're safer in town than in the country.

And Keith:

We had a neighbour in Tooting who – spasmodically – homophobically harassed me. He'd shout 'homosexual pervert' when he saw me. I put up with it for a year, then went to the police. They gave him a warning but it didn't stop him. He then got arrested and bound over, then stopped doing it. It was a really awful feeling – I didn't want to go into the garden. Once or twice people have been threatening in straight pubs. I would avoid that kind of situation now – though because I'm older I probably wouldn't be seen as a sexual being. Cruising places attract queer bashers and people get attacked. Straight people do things in public – it's not fair that gay people can't.

Women, though, appear to be less fearful of outright homophobic violence. This reflects the greater invisibility of women's sexuality and of lesbian women in particular. Cath put it this way:

I think casual homophobia is more directed at men. I do have women friends who've had things shouted at them – but that's because they were much more obviously lesbians. I don't bother about how I dress when I go out. I'm not worried whether I look like a lesbian or not. These days, if you're out with a bunch of women, who's to know that you're not a bunch of WIers rather than lesbians.

CONCLUSION

As public space is evacuated by older people, the space relinquished becomes ever more threatening as their presence becomes more and more unfamiliar and strange. This reinforces older people's sense of diminished rights to public space. What we see here, as in the lives of children, is the burden of risk reduction forming part of the moral responsibility of urban citizens themselves (Rose 2000: 103), rather than urban initiatives to address the real or imagined fears that are articulated. Older people's exclusion from participation in politics, in making decisions over their own lives, renders the public realm alien and disconnected from their concerns. In this context, U3A and allotments are crucial sites for breaking down the privatisation and powerlessness of this growing section of the population.

Missing in many North European and American cities, in contrast to cities in South European countries and other parts of the world, are safe, car-free streets, squares and piazzas where older people and others can pass the time of day without the pressure of obligatory consumption.

Plate 6.4 Chatting in the central square, Montone, Umbria.

The narrative presented here is not, however, predominantly one of gloom. Though older people are typically ignored and marginalised, sometimes constrained in their access to formal public provision or daily street life, the city nevertheless offers countless sites and spaces of sociality and encounter, of delight and desire expressed in multifarious and serendipitous ways in public space.

7 Children's publics

> Sometimes the fragment of landscape thus transported into the present will detach itself in such isolation from all associations that it floats uncertainly upon my mind, like a flowering isle of Delos, and I am unable to say from what place, from what time – perhaps, quite simply, from which of my dreams – it comes. But it is pre-eminently as the deepest layer of my mental soil, as firm sites on which I may still build, that I regard the Méséglise and Guermantes 'ways'. It is because I used to think of certain things, of certain people, while I was roaming along them, that the things, the people which they taught me to know, and these alone, I still take seriously, still give me joy. . . . The 'Méséglise way' with its lilacs, its hawthorns, its cornflowers, its poppies, its apple trees, the 'Guermantes way' with its river full of tadpoles, its water-lilies, and its buttercups have constituted for me for all time the picture of the land in which I would fain pass my life.
>
> (Proust 1983: 217)

Recollections of childhood are typically imbued with a sense of place and space. Memories of events and experiences are suffused with, and in turn evoked by, sights, sounds, smells of this road, park, house, that alleyway, walk by the river, hidden cranny, this school yard, dark tunnel, railway and that playground, pool, farm. Idealised, romanticised or recalled with misery or sadness, places of childhood leave their traces in the public spaces of the city through which we pass. Nowhere is this more clear than in literary or autobiographical texts. These idealised landscapes, according to Ward (1978: 1) evoke a paradise lost and are frequently bound up with notions of rural family life. Raymond Williams (1973) (quoted in Ward 1978: 1) describes it thus:

> an idea of the country is an idea of childhood: not only the local memories, or the ideally shared communal memory, but the feel of childhood; of delighted absorption in our own world . . . but what is interesting now is that we have had enough stories and memories of urban childhoods to perceive the same patterns where these are now sites like the local corner shop or the rag and bone man.

For adults of the contemporary period, particularly those born before the latter part of the twentieth century, the dominant discourse of childhood is one of freedom of movement through the spaces of their neighbourhood, of lack of a sense of constraint, of walking to school, of playing with friends in secret places, of appropriating public space in hidden and magical ways. Though warned of strangers, for most people from these earlier generations, fear, avoidance and risk did not define life as a child. Rather childhood was figured as a time of innocence. In this there has been a sea change.

A growing discourse of fear and risk, what Sapolsky refers to as a 'collective mania of risk' (quoted in Hood *et al.* 2004: 3), has evolved which is pernicious in that it acts to produce the very circumstances of which it speaks. Parents' anxiety and fear limits children's movement and freedom (Valentine 1996; O'Brien *et al.* 2000). Yet as children are withdrawn from public space, through its emptiness and lack of use it becomes more threatening. The two main tropes deployed are the fear of attack or sexual assault, lodged at its most dramatic in the figure of the paedophile, and the fear of traffic. Owing to these perceived risks, children no longer walk to school, particularly alone, and sallies into the local area – to the shop, to the friend down the road or to the park – are restricted. That the massive increase in cars has led to an increased number of children killed or injured on the road cannot be disputed: in 2000, 150 children were killed and 25,000 injured by cars (Seaford 2001: 461). Yet the withdrawal of pedestrians from the street opens this space up for the car to roar at speed through cities with little to impede its progress.

Tudor (2003: 238) describes this fearfulness as a way of life in modern societies where 'our children are no longer allowed to walk to school, and the landscapes of fear we paint for them are populated not with trolls, wolves or wicked witches, but with paedophiles, satanic abusers and generically untrustworthy adults' (ibid: 239). The fear of sexual attack, produced in part by sensationalised media reports of paedophile rings and a number of highly publicised accounts of kidnapped children and sexual assaults – for example the Holly Wells and Jessica Chapman case in Britain – has profoundly altered the meaning of public spaces for children. A recent report (cited in Sieghart 2004) reveals that the number of abductions rose to 846 in 2002–2003 (a 45 per cent increase over the previous year), and that 55 per cent of these were by strangers, 23 per cent by a parent and 22 per cent by an adult known to the child). But the majority were unsuccessful – only 9 per cent were successful abductions by strangers – and according to the Home Office this increase can be explained by the fact that police are recording these crimes more readily. Whatever the level of real increase, the more important point is that a child is almost as likely to be attacked in the home as by a stranger in a public place.

In her survey in rural, metropolitan and non-metropolitan areas in the Northern UK of 8-11-year-olds' ability to negotiate public space alone safely, Valentine explored the notion that parental fears were squeezing children out of public space, thus 'producing it as "naturally" or "normally" an adult space' (1997: 83). Abduction was seen as the greatest danger by parents of primary school children (45 per cent), rather than traffic accidents (34 per cent) and drugs (9 per cent). These abductions were conceived as by strangers (63 per cent) rather than adults known to the child (1997: 70). They were most scared of public parks (60 per cent) as a site of potential snatching, followed by shopping centres (34 per cent), playgrounds (33 per cent) and outside school (6 per cent). Implicit here is the notion that children can't negotiate space alone, constituting them as parents' possessions and parents as gatekeepers protecting children from the dangers of the world (Seaford 2001: 461). One of the most important factors in increasing adult supervision is the emergence of 'neighbourhoods without neighbours, who hitherto were responsible for collective socialising of children and keeping an eye on them' (Furedi 1997: 127). In this world of strangers, he argues, it is difficult to trust.

This perception of danger is a gendered and racialised one. Hart (1979) found girls were seen as more vulnerable to danger than boys and were thus given less freedom of movement than boys. And though parents in Valentine's study (1977) were equally concerned for boys' and girls' safety there were gender differences in how boys and girls were represented: girls were seen as more sensible and responsible while boys were seen as more easily led and immature. For non-white children – in this instance Asian children – there was the added fear of racially motivated attacks (ibid: 71). Parents' inability to trust their children to manage their own safety in public spaces leads many to control their children's use of public space, setting boundaries for them (ibid: 72). Katz (1995: 3) deploys the notion of 'terror talk' for discourse which posits strangers as dangerous and heightens parents' and children's sense of risk in public space. Yet, as she points out, this overlooks the potential dangers of the private sphere where children (as well as women) are more likely to be subjected to some form of domestic violence rather than violence from an unknown stranger. Whatever parents' fears may be, much research has shown that children are very competent at managing their own personal safety (Katz 1995: 75) and many receive stranger danger warnings from an early age and gain a strong sense of invulnerability (ibid: 78).

In this climate of fear of and withdrawal from outdoor public space, there has been a marked increase in the provision of formal spaces for play – such as indoor activity centres, climbing areas and so on – provided by local councils on the one hand, and on the other, by a growing commercial play sector. There is an interesting paradox here. Privatisation seems to engender a greater sense of freedom in public for parents and children. Blackford's (2004: 229) participant observation research compared mothers' behaviour in a public neighbourhood playground with that of parents in a McDonald's Play Place in the Bay Area of California. She found that commercial play spaces actually relieved both children and parents from surveillance regimes, since the places are contained and safe. In other words, mothers' vigilance is relaxed in these spaces. As a result children's own rules for play and social interaction were more autonomous and creative (ibid: 229). At the same time Blackford found an elaborate domestication of the playground space by mothers, who spread out domestic paraphernalia on benches across the park, 'bringing their domestic authority to organise their children's bodies and space' (ibid: 239), whereas in the McDonald's space there was a clearer separation of children's and adults' worlds (ibid: 242). She raises the question of whether children feel a unique cultural recognition in this kind of quasi public space and concludes that: 'The growth of commercial play spaces in urban and semi-urban areas seems to signify the historical culmination of middle-class efforts to contain free play' (ibid: 244).

Degradation of the urban landscape is another route to the marginalisation of children from public space – particularly poorer children. Elsley's study of children of 10–14 years old in an urban regeneration area in Edinburgh found that the concentration of graffiti, urban crime and vandalism in areas of urban deprivation and disadvantage impacted on children there (2004: 156). Their experience was different from that of children from wealthier areas who could more easily play at home, and they were thus likely to be more vulnerable in public space. The children in this study had a longer list of informal places that they liked than of formal places designated for them. These included streets, shops – particularly the local shopping centre, their out-of-school group, and wild areas like cornfields, some ruins and some local woods. The places they disliked were often marked by unpleasant events that had taken place there, such as being threatened by older young people; or else they were places that were vandalised or unpleasant to look at or unsafe, such as burnt-out areas or places with rubbish in them (ibid: 158).

The picture, though, is not a homogeneous one. In many different cities and cultures – particularly those outside the richer Western countries – children work from an early age or live in neighbourhoods, townships and settlements where networks, patterns of interaction and surveillance operate as a form of protection for children who are thus free to roam more widely. In 1977 Lynch and others investigated how 13–14 year olds used unprogrammed spaces in Cracow, Melbourne, Mexico City, Salta (Argentina), Toluca (Mexico) and Warsaw. Though there were some general findings across these localities, such as the fact that the children talked little about school or their private yards, more commonly mentioning the street, their own room, sports facilities, wastelands and the city centre, there were also striking differences. The Australian children were the most mobile. Even at that time the common barriers to movement were personal fear and traffic, as well as lack of spatial knowledge, the cost of public transport and – for girls – parental controls. The children in Salta and Melbourne constantly mentioned boredom. There were clear gender differences in representations of space in Ecatepec (Lynch 1977: 44), with boys mainly representing their environment as a map of streets and blocks, and girls tending to draw shops, parks and green areas. Consistent themes in their utopian visions were trees, friends, lack of traffic, cleanliness, small size and quiet.

Children's experience is also heterogeneous spatially within Western countries. A comparative study of the UK and Germany revealed that adults in Germany were far more likely to let children out on their own and there was an expectation that other adults would keep an eye on them (Furedi 1997: 128). O'Brien *et al.* (2000), in a study of children living in two contrasting urban environments, London and a lower-density new town, found that there were significant variations in how contemporary children use their public spaces. Children's freedom to move around their neighbourhood was greatest in the new town and in line with other studies girls and ethnic-minority children were more restricted in their use of urban space. Observing eight playgrounds in Amsterdam, Karsten (2003) found that playground participation and activities were clearly structured by gender, while the specific physical and symbolic landscapes further reinforced this binary divide.

Race and gender clearly fragment any unitary narrative as to children's relationship to the public realm and public space. But even more significant is the place of class and income. It is now relatively widely accepted that children are as affected by socio-economic circumstances as adults (Elsley 2004: 155). For children who are poorer the street takes on a greater importance since there is less access to expensive leisure activities and centres (ibid: 156). In contrast, it is argued that children in wealthier communities are subjected to lives that are rigidly organised by adults with little opportunity for playing and socialising freely with their peers (Riggio 2002: 46). Such a contention is supported by Hart's research (2002) which revealed that children from lower-income urban neighbourhoods played outside close to home more freely than more middle-class kids. As a result, he suggests (ibid: 135) that as cities develop in the manner typical of the West there is a tendency for children to be increasingly contained. For Hart children's play should be a priority for city governments because 'first, play is important to children's development, and second, free play in public space is important for the development of civil society and hence for democracy' (ibid: 136). Since their development at the beginning of the twentieth century, playgrounds for children have been less popular than planners anticipated. Most people remember nostalgically not structured spaces but the creative and secret spaces of their childhood which they could adapt and transform imaginatively for their play, and where they could gain some sense of control. These are the disordered spaces which disrupt the 'striated fabric' in the modern cities of adults (Cloke and Jones 2005: 313). Hart argues that the establishment of playgrounds has amounted to a form of spatial segregation of children where they have little participation in the making of their own environment. In his view (2002: 137) children need opportunities outside the class room for socialising with children from different classes, cultures and ages so as to learn to co-operate and form relationships with them. This is why public space is so fundamental to civil society and democracy.

Ward (1978) also highlights class differences in his study of children and the city. Drawing on the Newsons' (1976) study he argues that working-class children, especially boys, saw the outdoors as significantly more important than middle-class kids (Ward 1978: 43). Revealed in this study also was the more limited spatial experience of working-class children surveyed in Handsworth, Birmingham, many of whom had never been to the city centre (ibid: 44). An interesting difference emerged between white and black children: many of the black children played indoors, seeing home as a haven from the potential threats outside. Yet these are the very children whose homes are likely to have limited domestic space for play.

Humphries *et al.* (1988) similarly emphasise a division which has been increasing since the mid-1980s between affluent and poor children in urban areas, where for the poorer children childhood remains short and brutal. This story is repeated the world over: extremes of starvation and violence sometimes mean that any consideration of public space is a luxury beyond most children's wildest imagination. Nevertheless, that too is only one story. There are many stories of tactics of resistance, spontaneous delight and acts of creativity in the harshest environments. Beazley's (2000: 484) study of street boys in Yogyakarta, for example, found that despite being marginalised from public spaces through verbal abuse, evictions, arrests and beatings, these children contested their exclusion through appropriating specific public spaces in the city and constructing a network of entwined spaces for their everyday survival and pleasure.

Another prevalent shift is the endless pressure on parents, from TV and other forms of advertising, to buy new commodities for their children – a pressure that more vulnerable and stressed parents, such as unemployed younger women on their own, may find hard to resist. Thus for many children, as Ward (1978: 125) points out, 'modern life in fact exposes the young to the cornucopia of consumer desires while progressively denying them the means of gratifying these expensive wants except through the munificence of parents'. But for the more privileged others, childhood is a period in which they are protected and indulged and where they have access to a whole host of institutions, facilities, shops, consumption sites and play spaces to satisfy their needs, living a life which would have been unimaginable a century ago (ibid).

CHILDREN AND PUBLIC SPACE: A CASE STUDY

This case study was set up to explore the different experiences of public space of children in two localities of London, one relatively well resourced with public amenities, the other with a low level of local public amenities. The aim was to explore both the impact of spatial differences on children's sense of public space and the continuing significance of class as a key social division in children's lives in the city. The children were selected from classes of 10-year-olds in two primary schools in inner London, one north of the river, one south. Charles Avenue School is located on the borders of Brixton and has an intake of 530 children – more than most primary schools. The school is situated in a disadvantaged locality, as is illustrated by the fact that 283 children (53 per cent) are entitled to free school meals. This is higher than the national average, as is the number of special needs children (73). In 2004, 254 children were from heritages other than English, the main home languages apart from English being Yoruba, Portuguese, Somali and Spanish. At the last OFSTED inspection 226 children were defined as being at an early stage of English acquisition, with many speaking no English before attending the school. Six were the children of travellers, and there were several Romany asylum seekers. Many of the levels of attainment were much lower than those normally found in children of an equivalent age elsewhere. The school has a high turnover of children and below-average attendance rates.

Rutland is a voluntary-aided Church of England school where preference is given to applicants who are churchgoers, and where Christian values are taught.

There are 257 pupils in the school, of whom a third (above the national average) receive free school meals. There is also an above-average ethnic mix in the school, with 40 per cent speaking English as an additional language. The main languages spoken at home are Bengali, Arabic, Albanian and Portuguese, and another 21 languages are also spoken. Twenty-nine of the children are from Kosovan or Albanian refugee families. One-quarter of the pupils are recognised as having special needs and a wide range of emotional and behavioural problems and difficulties was identified in the OFSTED (2002) report. In 2000 Rutland was designated as a school for children with disabilities. Despite this concentration of disadvantage, the school also has a fair number of middle-class children from the surrounding gentrified areas. The school has an ethos of caring and a strong emphasis on moral and cultural development and on the involvement of parents in the school's activities.

The two localities are also very different in many respects. Though both contain a mix of public and private housing, the square mile or so around Rutland School contains many playgrounds and parks, notably Hampstead Heath half a mile away, a city farm, an after-school club, a new sports centre called Talacre (sometimes referred to by the children as Treetops, the name of the younger children's climbing area), a good children's bookshop, a public library and two swimming pools. The locality around Charles Avenue is far less resourced with facilities for children, although some of the estates do have open play areas and play equipment, and Brixton market and the Ritzy Cinema are close by. There is a higher proportion of gentrified terrace housing in the immediate vicinity of Rutland School than close to Charles Avenue School.

Six of Rutland School's participants in the case study were boys, of whom four were black British (three Afro-Caribbean, one African); nine were girls, of whom three were black British (one African, two Afro-Caribbean), and six were white, including one Portuguese. Of the Charles Avenue participants, five were boys, of whom two were white and three were black British (two African and one Afro-Caribbean) and nine were girls of whom three were white British (one was Roma) and six were black British (four Caribbean, two African). In small groups the children were asked to draw pictures of the places they liked, the places they disliked and the places they visited most often. The children were also given disposable cameras to take home and were asked to take photographs of the everyday public spaces they inhabited. During the process of drawing, and after the photographs were developed, focus group interviews were conducted where the children were asked to elaborate on the pictures they had drawn and the photographs they had taken. What emerged were striking differences between the children's sense of public space in the two localities, and between the children in each of the schools, this difference primarily registered on class lines. These differences were also inflected by race/ethnicity and gender, though seemingly much less significantly.

As pointed out earlier, the majority of the children at Charles Avenue School were from poor backgrounds, and many were not white. The locality is also relatively run down despite large swathes of gentrification, and is marked by a lack of resources and amenities when compared to Rutland. In Charles Avenue a very distinct imagination and experience of place thus emerged which can be organised around three tropes: circumscribed personal geographies, an elision of the notion of public space with spaces of consumption, fear of violence and a prevailing sense of emptiness or space emptied of life and vibrancy.

Let us look at each of these themes through the children's drawings, photographs and interviews. The pictures from the Charles Avenue children are often badly drawn and contain spelling mistakes. When asked to draw places they liked (some drew more than one image, hence there are more images than children), the children drew the Brixton Odeon Cinema (seven), the park (one), the zoo, the street where they lived (three), the Ritzy Cinema (two), KFC (six), Woolworth's (four), JD Sports (two), Foot Locker (five – 'Half Price! On Sale!') a bus, the recreation centre (three), Game Parlour (two), Mega Bowl, the One Pound Shop, Mr Cheap, a funfair (three), the London Eye, a beach (two), McDonald's (fifteen), a hairdresser's, a sweet shop, a football club, Tesco (two), Shopping City Wood Green, Next, Nandos, Peppermint, school (one white girl), New Look, the Dome (two), Dallas.

There was a similar range of places which the children visited most often: McDonald's (sixteen), the Ritzy Cinema (four), Peppermint, Oxfam, a local shop or cola shop (six), the park (two), New Look (three), football (two), KFC (six), the recreation centre (two), Tesco (five), Londis, Woolworth's (two), JD Sports (two), Foot Locker (three), a fish monger ('fish munjer'), Iceland, a cinema, ASDA, a train station, church (Dorcas), a sweet shop, a street, a swimming centre, Costcutter, Clare's Accessories, Lolly Pop Boutique.

Figures 7.1 a–c Charles Avenue Schoolchildren's drawings of places most liked.

a Foot Locker and KFC.

Figure 7.1 (continued)

b The funfair.

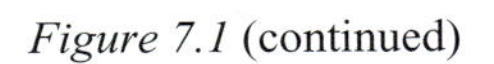

Figure 7.1 (continued)

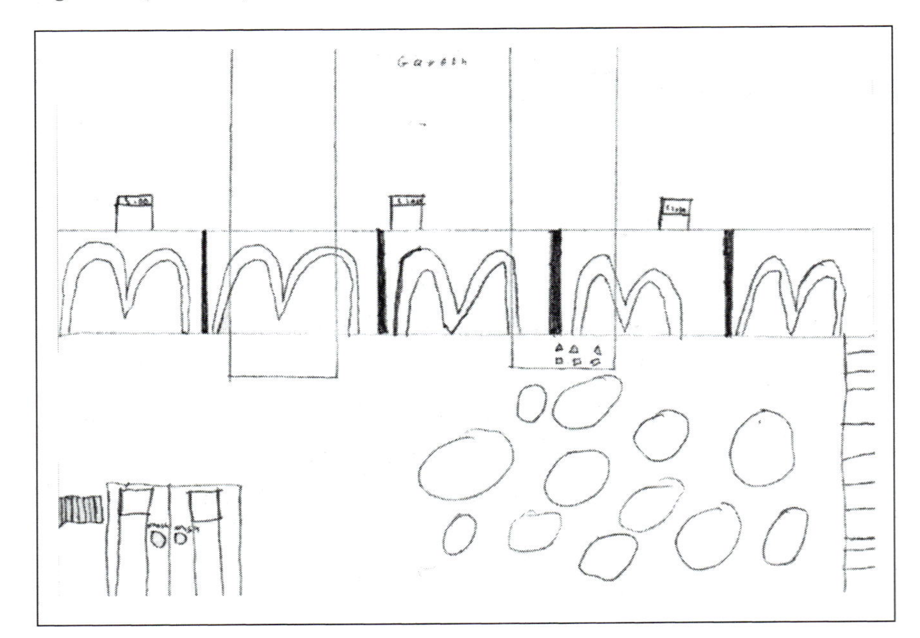

c McDonald's.

Figures 7.2 a–c Charles Avenue Schoolchildren's drawings of places visited most often.

a McDonald's.

Figure 7.2 (continued)

b Woolworths and New Look.

c The fishmonger.

The pictures of the places the children disliked were the least well drawn and detailed: the train track, a meat shop, a fish shop (four), a chicken shop (two), a carpet shop (two), the One Pound Shop, Tesco (three), litter bins (three), Marks and Spencer, Foot Locker, JD Sports, a laundrette, a Chinese fish shop, Clark's (two), the park, Oxfam, the dentist's, the optician's, Speedy Noodles (two), Iceland, New Look, the Odeon Cinema, the high street (two), nothing, KFC (three), Sainsbury's (two), a cinema, a game place, an Indian takeaway, baby centre. Three drew nothing and one was a very limited drawing.

The themes running through the children's pictures at Rutland offered clear contrasts: there was a prevalence of publicly provided spaces, the microgeographies were less circumscribed, and the sites of low-income commodity and food consumption were far less prevalent.

The pictures were also clearer and more detailed with fewer spelling mistakes. The places they described liking were as follows: the public swimming baths (two), the library ('I love it', said the British African kid who had drawn it), 'my cousin's Swiss villa', the church, the cinema (two), the after-school club, Safeway, 'my road' (three), football (four including three girls), Mile End climbing wall, Spitalfields market ('my dad has a stall there'), Canary Wharf, the park (six), Cantelowes Park (two), Sobell Leisure Centre, Somerset House, birthday parties (three girls), a fair (three), the Dome (two), a restaurant, a sweet shop, Cool Waters in California.

The sites that the children visited most often were: the city farm, church (two), 'my street', the swimming pool, the after-school community club (two), 'play out in the street', the local sports centre (three), 'my Aunty's place', school, library, 'my Nan's place', a local shop (two), basket ball, football.

Figures 7.3 a–c Charles Avenue Schoolchildren's drawings of places they disliked.

a The One Pound Shop and Marks and Spencer.

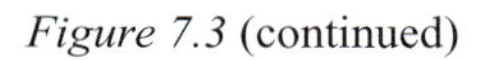
Figure 7.3 (continued)

b A fish shop and a chicken shop.

Figure 7.3 (continued)

c The train track.

Figures 7.4 a–c Rutland Schoolchildren's drawings of places most liked.

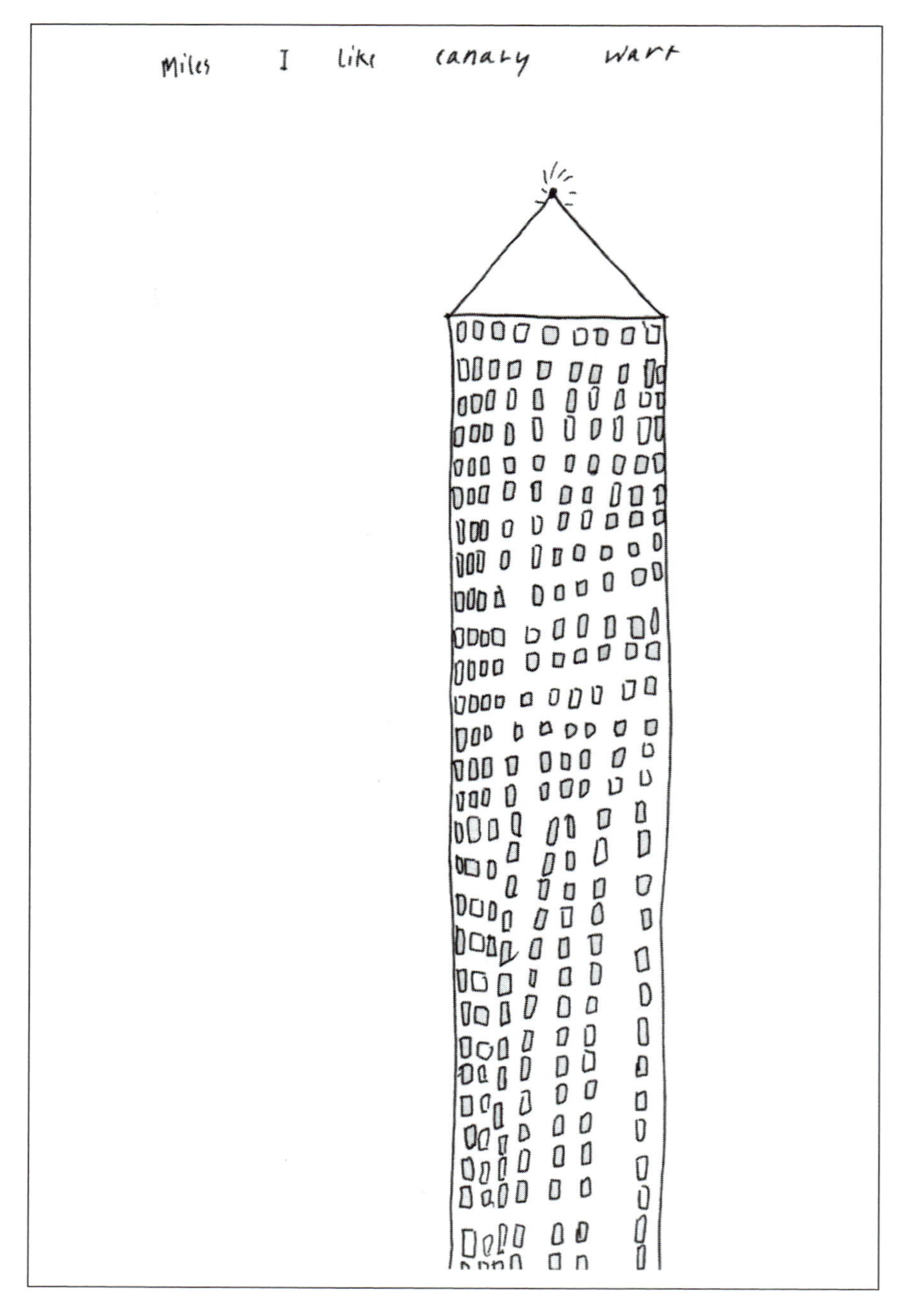

a Canary Wharf.

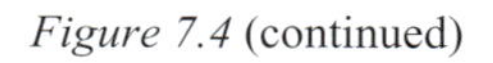

Figure 7.4 (continued)

b The swimming pool.

Figure 7.4 (continued)

c The Talacre community centre.

Figures 7.5 a–c Rutland Schoolchildren's drawings of places visited most often.

a The NW5 Play Project.

Figure 7.5 (continued)

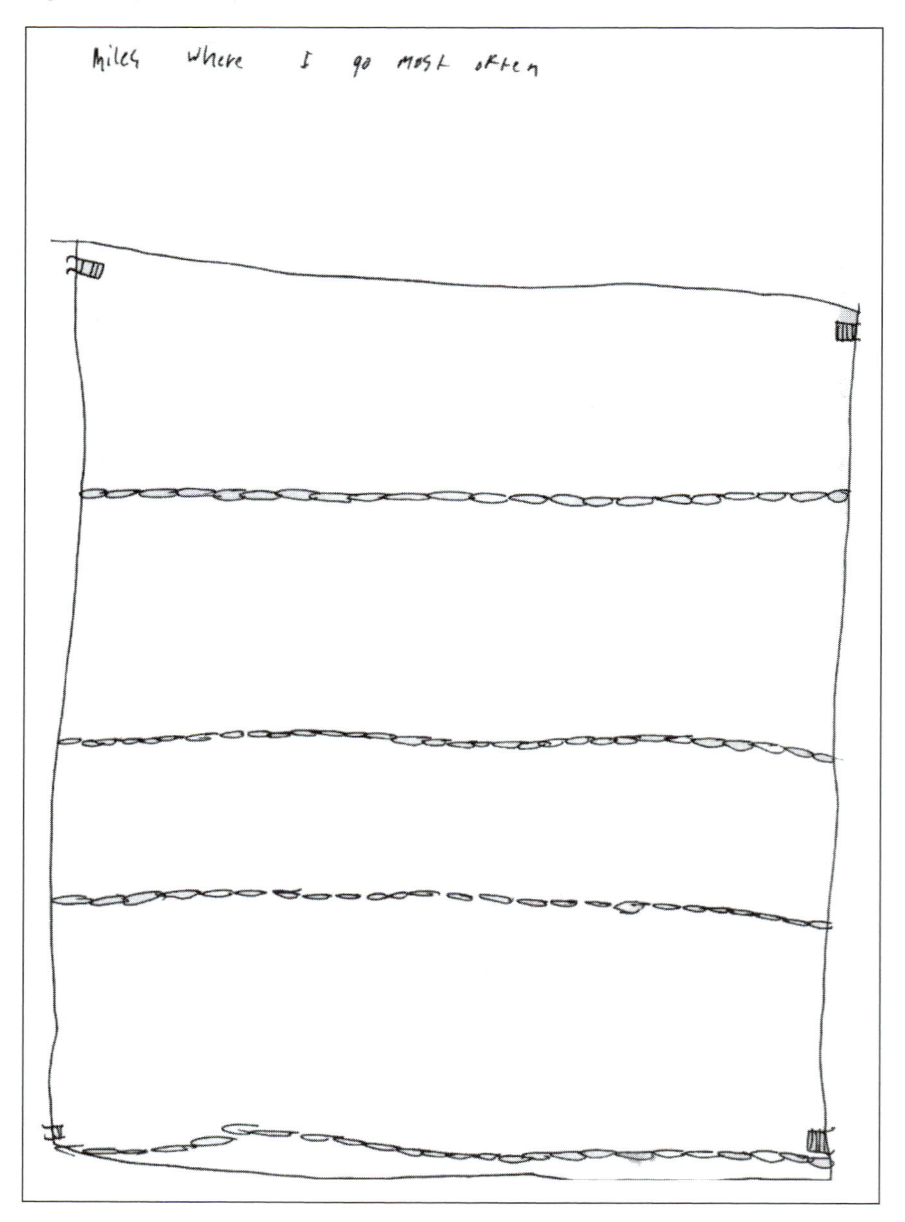

b The swimming pool.

Figure 7.5 (continued)

c ‘Play out in the street’.

Figures 7.6 a–c Rutland Schoolchildren’s drawings of places disliked.

a The supermarket.

Figure 7.6 (continued)

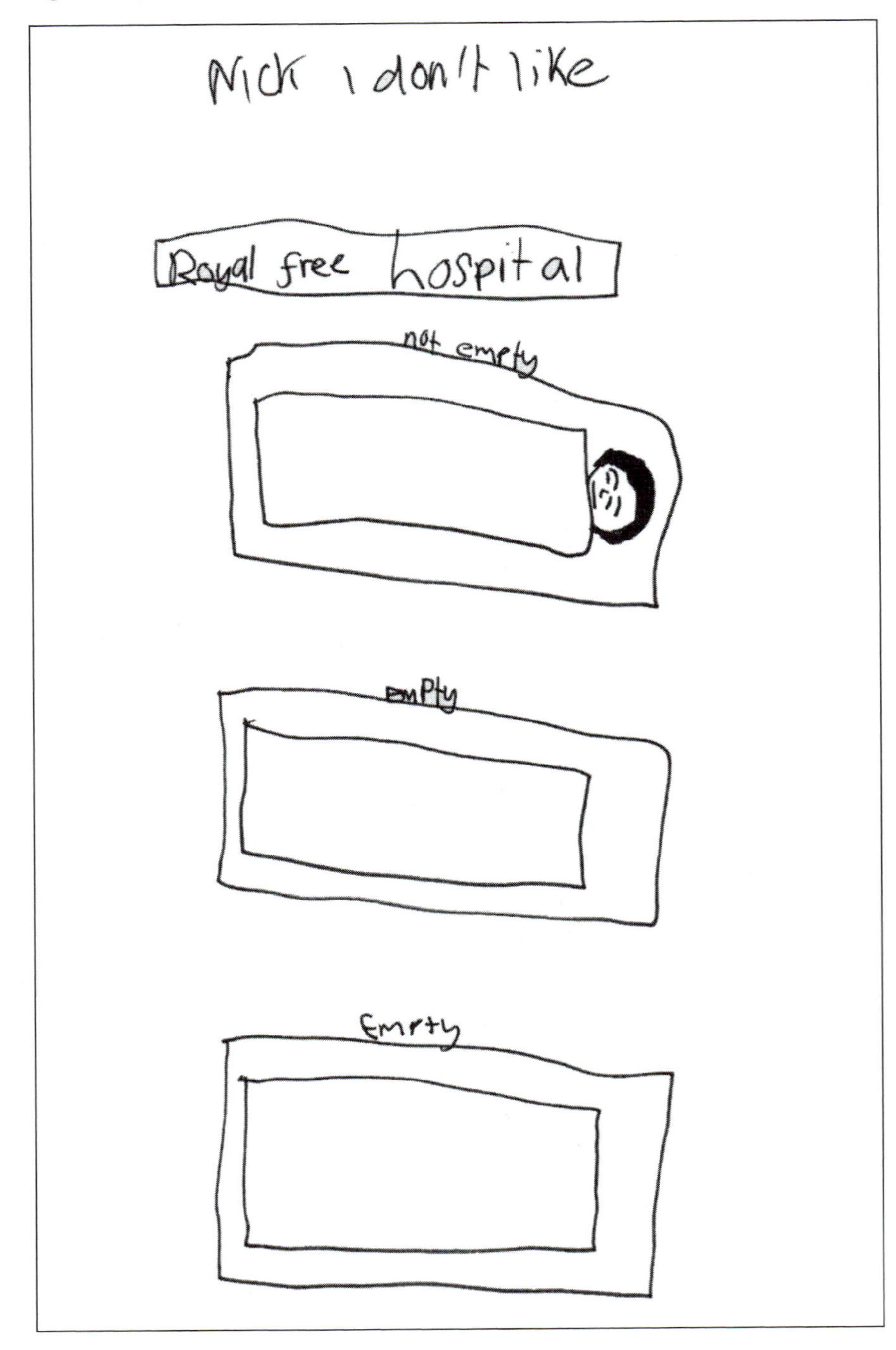

b The Royal Free Hospital.

Figure 7.6 (continued)

c McDonald's.

The pictures of the places that the children disliked were discussed and drawn with more noticeable affect. For example one black boy drew 'our neighbours at no 2 and 4' whom he describes as hostile and racist, three other black boys drew Hampstead Heath where they described feeling out of place or harassed. The other pictures were of the local sports centre, school (three), Camjam (a children's music workshop), McDonald's, the dentist, Starbucks, 'my aunt's flat', the, Royal Free Hospital, empty parks, dustbins (three), Safeways (two), boys (two), Oddbins, the supermarket, Wickses DIY shop.

The children's choice of sites to photograph also offered very clear contrasts (see Table 7.1 opposite).

Looking through the photos taken by the Brixton children, I was struck by a prevailing sense of emptiness. Many of the landscapes were barren and bleak. Empty walkways through estates, cars on the street, the odd rather run-down-looking local shop, rickety play equipment, litter – these were the images that dominated. They were by no stretch of the imagination spaces of easy sociality. A brief summary of some of the children's photographs will give more depth to this impression. Ken – a black child – took the following photographs: a pavement, a concrete slide (four photos), two children at a petrol station, a kid by a wall and another by a garage (two of these), an estate walkway, a concrete football area where black children were playing (four of these), litter. Lin's photos were of a block of flats, an empty pavement, McDonald's, the Ritzy Cinema, some girls in head scarves, a shop, the bus stop, dustbins, Londis (two of these). Sarah's pictures were of McDonald's, an empty part of the estate, a chemist's shop, a street with a kid in it (two of these), the health centre, a shop and the shop owner (Muslim), the bicycle park of a shopping centre, an empty concrete football pitch, Costcutter, flats covered in scaffolding (two of these), an empty walkway, children in a tunnel on the estate. Tanya's photos were of a petrol station, a road with cars and no people in it, two children behind a fence on a housing estate, a supermarket, ASDA, McDonald's, a row of shops, inside the corner shop, a shopping street, the estate (two), three girls in a playground, an empty slide in an empty playground, two children and a baby by the road between some cars, the bus stop, a low-cost vegetable shop, a Brixton monument viewed from the bus. Annal's photographs were of a bus, an empty playing area (three), an empty park (four), and some stones in the park.

The Rutland children's photographs tended to be more cheerful, full of people, and diverse in their range, and less bleak. There were more pictures of public spaces provided by the council – such as Talacre and the after-school club – and Hampstead Heath, the photographs were more populated, there were fewer pictures of shops selling cheap products, more photographs of places that were not in the immediate locality and overall more variation. These profiles provide some illustration. A white middle-class girl, Emma's photographs were of: a telephone box, Rutland station, the public library, Nandos, McDonald's (because she hated it as the 'food is bad for you'), Victoria Park in Hackney (four) – including Charlotte on a bicycle and a fountain – her school (four), a parking meter, her street of very gentrified terrace houses (three), men digging up the road, the

Table 7.1 A comparison between photographs of local everyday spaces taken by children from the two schools

	Charles boys	*Charles girls*	*Rutland boys*	*Rutland girls*
Estate, empty	xxxx	xxxxxxxxxxx	xx	xxx
Walkway	xx	xx		x
Street or pavement, empty or car lined	xxxxxxxx	x	xxxxx	
Street or estate with kids	xxxxxxx			xxxxxxxxx
Play area, empty	xxxxxx	xxxxxxxxxxxxxxx xx		
Park, empty	xx	xxxxxxxxxxxxxxx		xxx
Park with kids			xxxxxx	x
After-school club			xxxxxxxxxxxxxx xxxxxxxxxx	
Sports centre			xxxx	xxxxxxxxx
Pool			xxxx	xxxx
Cinema			xxx	
Football pitch		x		
Play area with kids	x	xxxxxxxxx		xxxxxx
Library/book shop			x	xxxx
Cars, garages, car park		xxxxxx		x
Shopping street	xxxxxxx	xx		x
Shop, local	xxxxx	xxxxxxxxx	xxxxx	xxxxxxxxx
Supermarket	xx	xxxxxxxxxxx	xxxx	xxxxxxxxxxxxxx xxx
Litter or bins or graffiti	xx	xxxxxxxxxx		xxxxx
Public facility (e.g. health centre)	xx	xxxxx	xx	xxxxxx
Bus/tube station		xxx	xxxx	xxxx
Restaurant, café, McDonald's	x	xxxxxx	xxx	xxxxxxxxxxxxxx
Pub	xx			x
Church		xx	xxx	xxxxx
Total number of children in group	5 (2 white, 3 black)	8 (3 white, 4 black, 1 Roma)	6 (5 black, 1 white)	9 (3 black, 5 white, 1 ?)

Notes:
The majority of streets photographed in South London are shopping streets. The ones in North London are usually residential streets. Domestic spaces, internal and external, e.g. the garden, are not recorded in these tables.

Plates 7.1 a–f Children's photographs of public space in their locality

a A South London council estate.

b McDonald's, Brixton, London.

Plate 7.1 (continued)

c Argos, Brixton, London.

d Camden Lock, London.

Plate 7.1 (continued)

e Kentish Town City Farm, London.

Plate 7.1 (continued)

f NW5 Play Project, London.

dry cleaner's, a DIY shop. A black African boy, Jonas' photographs were of: the local post office, a local shop (two) the chemist's, his house, the public library, the after-school club (two), a shopping street, an enclosed public market space, the station, a bus stop, a church, Rutland Sports Centre and Talacre community centre. James (a white boy) photographed Rutland Sports Centre, Rutland station, a church, his street (two), the local shop, a trendy local café (two), his house, a newsagent, a Tai Chi centre (two) and Hampstead Heath (five). Finally Alison's photos included Somerset House, the London Eye and other parts of Central London.

Despite the overlaps in the sites selected for the drawings and photographs there were distinct differences within the schools and between the schools; these were reflected also in the group discussions. There were several key themes here. First, the microgeographies of the poorer children appeared to be far more circumscribed than those of the more middle-class children (all at Rutland), who traversed, and were cognisant of, a broader range of space. The following comments provide an illustration.

SUSAN (white middle class): I like Spitalfields market . . . (A black child interjects: 'Where is that? I have never been there.') I like going to Spitalfields market cos my Dad has a stall there every Sunday and I like spending time with my Dad, cos I don't get to see much of him and there are some nice things there . . . and also I like Mile End climbing wall.

JAMES (white middle class): I like Canary Wharf. Me and my Dad ride there along the canal on Sundays on our bikes and have ice cream when we get there.

DEBBIE (black girl): I like the Sobell Centre because I can do loads of different things, I can go to the pirate's place and I like Somerset House where I can go ice skating – I just like it.

EVA (white girl): And I like Cool Waters in America cos it's got slides and pools and I get to go there every year.

JONAS (black African boy): And I like my cousin's villa in Switzerland cos it's got a big attic and a basement and lot of space to play.

LIZ (white girl): I like the wheel cos I can see all of London and like playing football cos it's fun.

The poorer children, and the black children in particular, in both schools, described a more local environment:

EVIE (black twin): I like my street – I play with my friends every day. I like riding my bike down the street. I like to play football.

ALISON (white girl): I like going over the road to the flats cos I can play football there and my Mum thinks I can protect myself . . . she lets me out on the estate as long as I'm with my friends, if someone tried to kidnap me my friends would look after me. Easy to make friends when you live on the estate, when you come back from school you can just come and play outside. I like the park to play on the swings and I like the fair and birthday parties.

A striking difference between the children of the two schools was that the Brixton children (none of whom were from middle-class families) described liking supermarkets, shops and cheap eating places – particularly McDonald's and KFC – with much greater frequency:

AMY (black girl): I like Safeways – even though I don't like to shop . . . I just go anyway. I like riding the trolleys.

A discussion between two of the children developed as follows: 'I go to shops by myself, when I'm 11 I'll go to school by myself. I'm allowed to go and get sweeties and things for my Mum. I enjoy that. I am getting my own mobile when I am 12 for emergencies. I'll be able to go out more by myself with a mobile phone.'

'I go to Sainsbury's for my Mum . . . I try and fiddle a bit of money for myself.'

'I go up to Londis to top up the electricity card. The other one – I do that at Iceland.'

'My Mum lets me walk around Woolworth's . . . it's all right in there . . . I like the toys.'

'When my Mum's drunk we get taxis.'

JOSEPH (black boy): Londis – I come there with my Mum and by myself . . . I like it when I can buy whatever I like – but when she tells me what to buy I don't like it . . . it's boring.

A further apparent difference between the schools was that the Rutland children, even those from poorer backgrounds, took greater advantage of and enjoyed local amenities and resources. This reflected both a higher level of provision, on the one hand, but also a greater use of existing resources, on the other.

EMMA (white girl): I drew the swimming pool cos I like going swimming and I like seeing the progress I'm making. And I drew the park cos I like animals and plants and I like the park.

GEORGIA (black girl): I drew Warners Cinema as I like watching films and I get to see my friends and after that I normally go somewhere else. It's the one in Finchley and there's other things around there like bowling.

RAVEN (black twin): I like my skateboard and I like City Farm cos Calvin works there and I go there most of the time. I like Cantelowes Park cos I take my skateboard there . . . but I don't do it much cos there's teenagers there.

JESS (black boy): I like Cantelowes Park cos I take my remote control car there. And I like the City Farm cos I like brushing down the animals.

JONAS (black boy): I like club cos it's nice and you can play there and there is lots to do. I like church cos after church there is this big party. And I like the library cos I really like reading and I like the swimming pool cos I like swimming and that is the way I exercise and I like the cinema cos I like watching films.

Jonas' comments about the library cropped up in a number of comments from the Rutland children. This confirms the importance of libraries revealed in other research. 111 million children's books were borrowed from public libraries in the UK in 1995/1996 (Elkin and Kinnell 2000: xii) and the numbers are increasing. Elkin and Kinnell make the point that the public library is the only statutory local government service available to children from babyhood to adolescence. They also found that the proportion of children from black and minority groups using libraries was much higher than in the population as a whole (ibid: xiv).

EVA: I like Talacre as I get to play football and I like Treetops.

BELLE (white girl): I like the park, birthday parties, the Dome, the Wheel, the football pitch, the swings at the park and the slide. I like monkey bars and playing basket ball.

SARAH (black girl): I like the doctor surgery – it is a public place. It's cosy, it's quiet, I like it.

In the places cited as ones they disliked there were common themes across the localities: a sense of boredom which at the same time seemed to mask a sense of powerlessness:

LIZ: I drew Sally's house – my Mum's friend – I hate it, it's so boring. Last time we went there she made us sit on the stairs out in the hallway doing nothing while she had her hair done.

GEORGIA: I don't like empty parks cos say it's a rainy day and there is all this mud all over the grass and everywhere and it's so boring and there is not much to do there – just walking. I don't want to come there no more.

SASHA (mixed race): I don't like shopping cos it is so boring.

ALISON: I hate Wicks cos when my Dad starts looking at things there he takes ages and I don't like bins cos they are so smelly and they get on fire.

JENNY: I don't like Oddbins cos when my parents take me in there it takes ages for them to choose and it's boring. And I don't like bins cos they smell and I don't like boys. [One boy yells 'Sexist!']

CLARE (white girl): I don't like Safeways cos if your mum goes in there it takes so long and it's boring and I don't like boys and I don't like dustbins.

A number of the children, particularly girls, articulated a dislike of rubbish, litter and graffiti. In one of the Brixton children's accounts (Tania, black girl) this disgust is dramatically described:

> I used to go to this park but now all the dogs go there and there is poo everywhere. I don't go to the playground. My Mum says people smoke there and all the swings are broken . . . and they drink there . . . and drugs and there are syringes there. In the park behind my friend's house there was a needle. I said don't pick it up you never know what might be on it . . . could kill you.

Another issue depicted or mentioned in both schools was a sense of bullying or threat from other children or neighbours, particularly amongst the black kids. Eric again:

> I don't like school cos my parents come in on parents' night and then I get into trouble cos I talk too much and I don't like Rutland streets cos there are people who chase me. I like my street in Muswell Hill.

JONAS: I don't like Flat 2 and Flat 4, I live in Flat 3, cos there is this guy who is racist there and a drug addict and alcoholic and there is a racist in Flat 2.

JAMES: I don't like Camjam cos there is this boy who bullies me and beats me up.

RAVEN: I don't like the Heath cos there is this guy who chases me all around.

Many of the children, particularly the girls, mentioned a fear of drunk men in the street, but violence figured far more dramatically in the stories of the Brixton group:

SARAH: I'm not allowed to go to Brixton market alone, there are too many druggies there. Unless I go with older kids. Only allowed out near my house. Mum afraid I'll be kidnapped.

ANOTHER BRIXTON CHILD: I like watching *Crimewatch* – it makes me scared cos I see Acre Lane and Brixton. Two girls got raped opposite my house . . . But I like seeing what is going on everywhere . . . When I am with other people I feel safer . . . it can be scary. You hear about kidnapping. But I feel safe when I am with my friends' parents cos they can tell people to go away.

SARAH: I don't like the cab place – so I cross over the road . . . I could see there's guys in there drinking.

ANOTHER BRIXTON CHILD: Opposite the town hall there is this bus stop. There is this man who sits there. Has been sitting there for 8–9 years and he talks to me and also scares me. Men drinking scare me . . . I don't often see women.

SARAH: People who take drugs too . . . and people with scars on their face . . . scary . . . if they did it to themselves like crazy people they might do it to me. People who look posh, dressed well, no scars on themselves and hair all nice . . . they don't scare me.

ANOTHER BRIXTON CHILD: It's really scary when people talk to you in the street.

A couple of children in both schools drew, or took photographs of, places associated with fear or pain:

EMMA: And I drew the dentist cos I am scared of going there.

TIM (black boy): I don't like hospital – it's scary and the nurses' and doctors' hands are cold.

STUART and SARAH: I like the doctor surgery – it is a public place. It's cosy, it's quiet I like it. I hate hospitals – give me a headache. They are scary.

Finally, a couple of the middle-class children's comments seemed to echo the opinions of their parents in ways reminiscent of the children in Ken Loach's film *7 plus 7* made in the 1960s:

EMMA: I don't like McDonald's cos there are all those horrible people and it makes you fat. All the people there eat fatty food and it is not proper food and it's an insult and you're hungry two minutes later (Two black boys in the group intervene to disagree) and God knows what is in there.

SUSAN: I hate Starbucks. There is this Starbucks near my brother's choir and it's really dirty and the water is disgusting. Once I took my friend there and she hadn't been before and it was really dirty.

Given the prevalence of adult memories of childhood which nostalgically return to the liminal, secret, in-between and mysterious spaces carved out in the ordered landscape of an adult world, supported in Borden's (2001) account of the spatial practices of young people skateboarding in the city, there is a notable absence in all these children's accounts of equivalent sites on the margins. Instead their focus, as we saw, is on public or commercial provision of space. Whether this reflects the relatively young age of the children interviewed, and hence a considerable restriction on their ability to 'roam free', as it were, or whether this reflects a more general shift away from these transformative and transgressive play practices, as spaces for their enactment are increasingly withdrawn, is impossible to assess. Their lack of representation, in the children's pictures or photos, was indeed a surprise.

CONCLUSION

What are the implications of these findings for contemporary urban life and policy? As Ariès (1962) famously pointed out: 'childhood is a modern invention', and indeed, it could be argued, a Westernised one. As we have seen, over the twentieth century the public realm for children has evolved as a space of fear and risk, of exclusion and segregation, and of privatisation and commercialisation. From the beginning to the end of the twentieth century children became increasingly regulated, subject to surveillance and contained in spaces specifically designed for their use, while spaces for spontaneous play and interaction, for rubbing along with unknown others, diminished. Thus, one of Ward's key arguments is that 'The failure of an urban environment can be measured in direct proportion to the number of playgrounds' (1978: 73), since in his view children will play anywhere, and park and playground designers have usurped children's creativity by forcing them into fixed places to play. Elsley (2004: 155) in a similar vein suggests that children generally occupy spaces within a world constructed by adults, outlawed from public space and contained within spaces specially designated for them like schools. Children's views as to the kinds of public spaces that work for them are largely ignored. In parallel there has been a domestication of childhood where the home/private space has become increasingly significant for children's play, resulting from fear of the streets and an increase in solitary play facilitated by the computer and video games (Sutton-Smith 1994: 137–139; Goldstein 1994). Children have thus become largely invisible (Matthews *et al.* 1999).

Yet this picture too is fractured and cross-cut by place and culture. As Jans points out, contemporary childhood is an ambivalent social phenomenon since children are seen as autonomous individuals as well as people in need of protection (2004: 27). In the context of the risk society of late modernity, characterised by individualisation and globalisation, there are new possibilities which are also more open to children (ibid: 28). One of these is the UN Convention on the Rights of the Child which, Jans (ibid: 290) suggests, was only made possible by globalisation and by the concept of supporting children as individuals with their own rights. Under the UN Convention on the Rights of the Child, children are seen as fellow citizens and there is a new vision of childhood. A related initiative is the International Child Friendly Cities Secretariat which was established at the Innocenti Research Centre, Florence, by UNICEF, UN-Habitat and the Italian government. This provides information and support to national governments and NGOs, children's groups and experts committed to making cities more child friendly and to fulfilling the rights of urban children. The Child Friendly Cities Movement represents a loose network of cities with governments committed to these principles. The Growing up in Cities project is an international research programme to help children in low-income urban areas to work with local officials to improve their areas.

Most of the research reported in a special issue of *Environment and Urbanisation* (Bartlett 2002) draws attention to children's powerful desire for inclusion in their communities. Many children describe feeling isolated where they live, excluded and threatened by traffic hazards, fear of violence, pollution or the absence of accessible public space. They also perceive a general lack of interest in their priorities (ibid: 4). Bartlett concludes that most cities are unfriendly places for children to live, reiterating Ward's (1978) point that playgrounds can undermine the capacity for free play in its diversity on the one hand, while also being an integral part of the urban fabric on the other. There is thus, she suggests, a danger that formal provision can be regarded as a substitute for real integration of child-friendly spaces into urban neighbourhoods – which is fundamental for the development of democratic civil society.

Many others have come to similar conclusions, variously stressing the need for children's citizenship, participation and inclusion in decision making, and a recognition of children as the 'neglected other' (Valentine 1997: 65). Elsley (2004) concludes her study by arguing for greater participation by children in the processes of area regeneration and for the influence of children's experience and views on public policy more generally. Various pieces of research suggest children are quite clear about what they desire and need. Gallagher's (2004) study of the 'Our Town' project, where children designed and built an intervention in their neighbourhood, found that children were conscious of the importance of inter-generational needs and interaction, safety, comfort and visual delight.

Certainly public space which is more child friendly is likely to better suit the needs of older people and ethnic and other minorities. Following Ward's (1978) notion that what is needed is a city where children can live in, and share, the same world as adults, the implication is that the city needs reshaping and redesigning so that children can refigure and transform the urban spaces they inhabit, and be integrated rather than being hived off into specially designated areas. Currently the world of the child is the sphere of the weak and socially manipulated. Children need opportunities to mix and mingle on the street (as Jacobs 1966 famously argued), breaking down strongly classified homogenous and ordered spaces where boundaries are clearly defined, and creating instead weakly classified spaces with permeable boundaries where mixing and diversity are enhanced (Sibley 1995). Living in the same public spaces as adults can be argued to serve yet another purpose. Thompson (1993), reviewing Carr *et al.*'s (1992) book on public space, argues that the emphasis on the public realm as a site of collective memory, cultural integration and environments for learning social relations

> reveals how well experiencing life in adult public space teaches children about their place in the human world. While witnessing the traditions and rituals of their own community, and encountering the strange ways of people unlike themselves, children grasp social nuances and build skills in human relationships and civic relatedness.

Finally, it is imperative that the diversity of children underpins debate about children's place in the city, implying a shift away from a notion of 'the unitary public child' (O'Brien *et al.* 2000). Disregard of the different ways children adapt, transform, imagine and live in the public spaces of cities will militate against any advances in urban policies designed to enhance participation for all children. This research highlighted the significance of class as crucially determining children's place in the public realm. Other differences, constructed by race, gender, place and culture, also disrupt any simple story of children's public space in the city.

8 The (dis)enchantments of urban encounters

Some concluding reflections

> It is frightening . . . how far we remain from mastering the sorts of allegiance, ethics, and action that might go with our complex and multiple belonging. . . . We should not and perhaps cannot accept the old cosmopolitan ideal of transcending the distinction between strangers and friends. Still, we all depend on what Blanche Du Bois calls the kindness of strangers. Less than kin or friendship but a good deal more than polite or innocent non relation, designating a field rather than an identity. . . .
>
> (Cheah and Robbins 1998: 3)

The very multiplicity of sites and subjects in public space which have made an appearance in this narrative underlines one of its conclusions – that there is no straightforward way of understanding urban encounters in public space. Temporality, spatiality and sociality intersect to construct spaces as enchanted to some and disenchanted to others, at different times of the day and night and in different socio-cultural contexts. Public space, and the encounters therein, can be no one thing. There are nevertheless some emergent themes and tentative reflections with which to conclude, not with the aim of formulating prescriptive policy and practice – though hopefully there are useful insights here – but more in the spirit of dialogue across difference and of throwing light on things that militate against rubbing along with others in public space.

PUBLIC/PRIVATE

First, the notion of the public is deeply implicated in its relation to the private and the boundary between the two. Notions of a public–private division have shaped cities since their inception, structuring and mirroring wider social relations and particular historical socio-cultural contexts. In post-industrial capitalist societies this division has demarcated the private sphere as primarily associated with the family, domesticity and privacy while the public sphere has been associated with employment, with formal institutional and political life. Urban analyses of the home (e.g. Saunders 1990) have pointed to its significance as a site of control, comfort and security – a haven in a heartless world. The public–private binary has long been subject to critique as artificial, notably coming under attack during the 1970s and 1980s, particularly from feminist theorists, sociologists and urbanists alike (e.g. Barrett 1980; Watson 1986; Hayden 1996), who argued that it reinforced patriarchal social relations and a particular family form where women are responsible for children and domestic life, and thus more marginalised from economic and political life. This picture has begun to change since the early 1990s as more and more women have entered further education and the work force, and social-cultural patterns of childrearing and family life – at least in cities – have begun to shift.

Nevertheless, the way the public–private division is understood remains a key part of how people live together in cities, determining to some extent the routines of daily life. As Madanipour (2003: 63) points out, the 'boundary between the public and the private, as any other boundary, is an expression of power that can subdivide space, give its subdivisions different meanings, expect the others to share these meanings by believing them'. Certainly the conflict over the eruv can be read in precisely this way, representing as it did a profound challenge to the public-universal/private-difference basis of modernity. In the context of the former, the eruv appears ridiculous, pointless and also impractical. In this sense the annoyance and resistance to it expressed by non-religious Jews may have less to do with fears of being tainted by association with an orthodoxy to which they do not subscribe than with their adherence to modernity and rationality. In the baths and ponds sites also, part of the story concerns what is appropriate to do in what spaces, which raises questions as to the extent the public can be contaminated by activities that are typically performed in private. Bodies and how these are performed in what spaces are central here. What rapidly becomes clear is that notions as to what is appropriate to do in public space are embedded in cultural discourses and practices which are normatively defined from a Western (specifically Anglo-Christian) perspective.

It is useful to consider Goffman here. In *Behavior in Public Places* (1963) and *Relations in Public: Microstudies of the Public Order* (1971) Erving Goffman provides a detailed account of social/public order, social relationships and public life. Goffman's concern is to examine the rules, norms and codes through which appropriate conduct is produced and exercised in public arenas and spaces. Central to his analysis is the complex interaction between private and public, and how different forms of behaviour are seen as acceptable in one and not another, thereby defining the very boundary between them. Underpinning his analysis (1963: 8) is the recognition that the distinction between acts that are approved and those that are deemed improper can only be made according to the judgement of a particular group. There can be no absolute form of arbitration (ibid) and what is proper in one situation may not be in another. So too in his work there is the recognition that multiple social realities can occur in the same place – mingling or conflict – though these may be temporally distinct.

As Goffman (ibid: 11–12) argues, norms supporting public order regulate face-to-face interaction. To be authorised to be present in a situation individuals must comport themselves in various ways; for example, they must not attract undue attention to themselves, must fit in and keep with the spirit or ethos of the situation. In describing street life he puts it thus (1971: 17):

> City streets, even in times which defame them, provide a setting where mutual trust is routinely displayed among strangers. Voluntary coordination of action is achieved in which each of the two parties has a conception of how matters ought to be handled between them, the two conceptions agree, each party believes that this agreement exists and each appreciates this knowledge about the agreement which is possessed by the other. In brief, structural prerequisites for rule by convention are found. Avoidance of collision is one example of the consequences.

Goffman's focus is on embodied communication and body idiom as a conventionalised and normative discourse (ibid: 34–35). He draws attention to the special mutuality of social interaction, how each person sees the other seeing him/her: 'Co presence renders persons uniquely accessible, available, and subject to one another. Public order, in its face-to-face aspects, has to do with the normative regulation of this accessibility' (ibid: 22). Geertz (1973: 448) similarly sees streets as essential for the enactment of rituals which provide an opportunity for a society to narrate a story about itself.

My argument here is that underlying some of the resistances to rubbing along in public, encountering others who are different, is a distaste towards others who behave in ways that are deemed inappropriate and unacceptable, often because they are designated as 'private', and that this produces and legitimates hostility in the self to others who are different. In a survey of a predominantly English and non-metropolitan group of 220 respondents contacted through the Mass Observation Survey in 2003, these conclusions were borne out. The following illustrate some of the comments made in response to a question as to what people would do in private but not in public, many of which are framed with reference to bodies and corporeality:

> Walk around in my night clothes or naked
> Swear
> Cough up phlegm (would do in a toilet in a public space)
> Pick my nose
> Dance uninhibitedly
> Make love (would do in a suitably secluded public area. Errr . . . a private or public area?)
>
> (H1745 f 52 London)

> Of course there are things I would do in private but not public – sex is an example. Urination is another. Any other human habits like picking one's teeth tend to be private. I can lose myself in my thoughts in private space, whereas I have to be 'on the ball' and thinking about issues like other

> people's feelings, appropriate behaviour, safety, etc., outside. My car feels fairly private although it obviously isn't in many ways.
>
> (R2862 f 44 Darwen, Lancashire)

> In public space I may impinge on other people, so I do not sing, make a loud noise, make lengthy calls on my mobile or take up a lot of space. At home I can sprawl, drink wine and dribble peach juice down my chin.
>
> (G2640 f 51 Hounslow)

Reflected in these comments are definite notions as to what should be exposed or not in the public realm. Arendt (1958: 72) makes this point clearly: 'The distinction between private and public realms, seen from the viewpoint of privacy rather than the body politic, equals the distinction between things that should be shown and things that should be hidden.' Here is another respondent:

> When you are in a public space you have to act with decorum and not behave in an attention-seeking way. You haven't to shout loudly. You have to dress sensibly and not in a revealing way. In private especially with one's family you can act daft and lark around if you want, crawling round on all fours with the babies and generally be more relaxed. I don't like kissing a member of my family in a public space not even my husband. My mother insisted on kissing me in a pub once and I was so embarrassed.
>
> Young people take quite the opposite viewpoint and shout and scream and behave in a hysterical manner in public. Some snog each other and even have sex in public places. Round here the young people are quite capable of deliberately walking in the middle of a busy road to wind up the drivers.
>
> (C1713 f 55 Preston)

> In the car I listen and sometimes sing along to music. I eat mints and sometimes fart. The latter I wouldn't do in a public place and I certainly wouldn't eat in the street, which goes back to my schooldays and fear of being reported eating in public wearing the school uniform! Mind you, on holiday *abroad* – not in the UK – I may eat in the street. Contradictory aren't I? The only common denominator is eating ice cream out of doors!
>
> In public spaces you are seen by others, and certain standards of behaviour used to be observed, which now are disregarded, and people behave worse in public than they do in private. I refer to public spitting and littering, both in the streets and on public transport. On the tubes fellow travellers appear to eat entire meals and then leave the packets with remains of food on the seats and floor. The smell is horrible and you have to wade through litter. I was brought up to take litter home with me if I could not place it in a bin. The food is always placed in carrier bags by the shop/take away and so it is easy to carry.
>
> (W2950 f 42 Tottenham, London)

Plate 8.1 Washing up and childcare on a residential street, Luang Prabang, Laos.

These comments are interesting in that they reveal the plasticity of the public/private divide, where sites like the car or the mobile phone used in public can create a sense of privacy – a private space – right in the midst of a public space. Similarly there are interstitial sites in buildings – the porch, the front step or the colonnade – which blur the private/public boundary.

> My husband and I row at home – but not in public. I don't often wear make-up or perfume at home but feel 'undressed' in public places without my 'face' on. At home I like to close all the curtains and feel locked-in to privacy, especially in winter. We dress more relaxed at home.
>
> (G226 f 62 Flyde coast, NW England)

What Goffman overlooks is the profound gendering of these cultural practices. Butler (1993, 1997) makes the point that sexed bodies are produced rather than pre-given outside social relations and discursive practice. Public spaces are crucial for the construction of gendered identities. In the Mass Observation accounts women were very reflexive about how they performed their femininity in public:

> I'd never comb my hair in public, or renew my makeup, or go out in my rollers. Like all the women on my father's side of the family, older and younger, I wouldn't go out of the house without a basic makeup on my face – not plastered on or obvious, but there. Having been to a very strict girls' grammar school, I still can't eat in the street, even though it would no longer mean a detention, and I don't like to see people wandering along munching pasties and sandwiches. When I smoked, I never smoked in the street or on buses, though I always smoked on long train journeys. In private, i.e. in my own home, I'm more likely to be dressed in scruff order or even, if I've poshed up, in what we refer to in the family as 'deshabills' . . . which means night things and a dressing gown and slippers.
>
> In private, I'd eat spaghetti!
>
> At home, I suppose I just switch off where, in the street, there's always a sense of having a duty to yourself to make the best of yourself, to walk with your shoulders back and, as my father used to say, 'look as if you belong to somebody'.
>
> (W633 f 61 Darlington, Co. Durham)
>
> Things one does only in private – mine would include singing and whistling, belching, coughing and sneezing immoderately, sunbathing in skimpy attire, scratching oneself and picking spots, sprawling on the furniture. As school-girls we were always told NEVER to be seen eating in the street, I've stuck to that rule all my life – even now I only feel comfortable licking an ice cream whilst strolling along the sea front, or sucking a cough sweet whilst out walking, any further eating would embarrass me. I don't like to see people eating hamburgers, hot dogs and papers of chips whilst walking around in public.
>
> Out in the street I like to project a particular image, walking assertively and hopefully looking confident and self-possessed – I would hate to appear as an elderly housewife shuffling along with my shopping bag so when setting out for town I carry a plastic carrier bag in my handbag for later use, and prefer to think I look more like a smartly dressed business woman or career girl (at 70? Such vanity!).
>
> (T2543 f 70 Dudley)

What is not revealed in the comments from these respondents, since they were predominantly white, is how public practices are inflected by race so that even in cities and spaces in Westernised countries people of ethnic or racial backgrounds other than the dominant cultural group may inhabit public space differently, and this non-conforming behaviour is greeted with hostility. Elijah Anderson (1990) provides powerful insights into the way street practices and street etiquette are highly racialised. The locality he considers is that of the Village in New York where potential and actual street crime and personal safety are central issues for most residents, mobilising particular spatial strategies in the street. More specifically black men, particularly younger working-class men, are assumed to be the greatest perpetrators of crime and are perceived as more threatening. In this environment people learn to develop a 'conception of self' (ibid: 335) which through various forms of street etiquette and behaviour provides them with some safety. Though there is an assumption that middle-class whites and blacks share similar concerns, the assumption, Anderson suggests, ultimately breaks down, since the 'experiences of a person with dark skin are very different from those of a white person, for several reasons' (ibid: 332). One illustration of his argument is in his notion of eye work. Here Anderson points out that blacks perceive whites as tense or hostile to them, as they receive very limited eye contact in comparison with the level of eye contact between white men. Black and white strangers meet each other's eyes for a few seconds before averting their gaze (ibid: 346). According to Anderson, this eye work maintains distance for safety and social purposes. Similarly, an intricate street ballet is implicated in passing behaviour on the street where people move away from each other to eliminate fear in public encounters. For example, Anderson has observed women crossing the street when they sense a black man approaching from behind. As a response, in spaces such as these a law-abiding streetwise black man might employ strategies to set the woman at her ease by crossing the street first to avoid encountering her at all (ibid: 345).

Once we shift our gaze to cities in other parts of the world where many different cultural practices are performed, public space and expectations around appropriate social and embodied practices are differently figured. For example, in Papua New Guinea, the Binandere people designate a space between private and public known as the *arapa*, which is a kind of wide street which:

> is a physical space [which] holds all people who reside in it as a single entity. Its members enter into a contractual obligation with one another, which requires that, if and when one of them is affected by something else, then another member is expected to act in consequence of it. For instance, in . . . a clash, the arapa is zealously guarded so that no outsider enters it.
>
> (Waiko 1992: 235)

This is the space from which people speak to others in the community, which designates a web of networks of relationships and obligations to others (ibid). My point here is in some sense an obvious one: different cultures have different

Plate 8.2 A street barber, Hanoi, Vietnam.
Photograph: Barbara Kirshenblatt Gimblett.

understandings of space and the kinds of embodied practices which are appropriate or not in public. If we look at the photographs taken in Hanoi in Vietnam and Luang Prabang in Laos, it is clear that the private/domestic and public/outdoor boundaries are completely differently figured. Eating and washing or caring for children in the street are part of daily life, as are other bodily practices like having one's hair cut.

The point, though, is that dominant, preconceived and fiercely held notions as to what is appropriate behaviour in public act to exclude and marginalise others, at the same time as providing fertile ground for disenchanted – even agonistic – urban encounters. This can operate across racial/ethnic differences as well as those of age, as we saw in the last two chapters, or across gender and sexual orientation, as we saw in Chapter 5. To prevent these boundaries being deployed as an exercise in power, there needs to be a possibility for them to be redrawn, and for a greater flexibility to allow dialogue to occur (Madanipour 2003: 63)

STRANGER DANGER

Fear of others who are different, unknown, and thus perceived as threatening, is a related source of constraint to living with difference. Kristeva (1991), as I have discussed, locates this fear and hostility to the other as deriving from the externalisation of unwanted aspects of the self onto the other. It is also a product sometimes of lack of exposure and familiarity such that encounters with different others are almost shocking. Reading the material from the Mass Observation Survey in 2003, I was struck by how limited many people's experience of those who are different to themselves actually is. Living in the inner city as I do, it is hard to imagine. Similarly, reading any text on cities in late modernity, where global processes have disrupted the dominance of whiteness and Englishness to a considerable extent, visiting other towns in the UK or vast areas of the states between the two coasts of the USA, or the towns of Europe away from the capitals, one is continually struck by the apparent lack of ethnic/racial differences amongst the inhabitants. Some of the quotes from the Mass Observation Survey sound remarkably naïve, illustrating some people's complete lack of experience of encounters with others different from themselves, which for some had occurred so many years ago and were so unusual that they deserved comment. Many were racking their brains to remember such encounters, finally coming up with descriptions of French exchanges and holidays, an arena of difference so mild as normally to warrant little sociological comment. For others being in hospital was their only experience of encountering difference. There was an overwhelming sense in the accounts that far from these respondents inhabiting a world where they encountered different others, their daily lives were marked by a homogeneous, not heterogeneous, form of sociality. This response is typical: 'The last time I met a group of people who were very different was a visit to our church by a German band' (P1009 f 64 Aberystwyth).

> I don't meet people who are very different from me as I lead a very quiet life and only mix with old friends and my relations. I suppose the only time I meet different people is when I go abroad on holiday.
>
> (C2579 f 58 Lowestoft)

> The people most unlike me I have ever met were Japanese, but having said that, in these days of mass communication, all peoples surely are more used to each other and have more understanding. The meeting wasn't different to meeting any group of people, except that I suppose we were a bit more keen not to cause any offence by doing the wrong thing. I would say we interacted well. We dutifully sat on the floor to eat, with our legs under us, and then to the side when they became tired and stiff. The Japanese simply got up and sat on chairs when they were tired! There were no problems with the Japanese we were introduced to, but two things I found a little hard to deal with in Japanese behaviour as a whole. Firstly, people would sometimes look at us with a deeply inscrutable expression, as if they were looking at a piece of nothing, as if they would/could never respond. Secondly, people stared quite openly and, worse still, would openly giggle and point. Fortunately we had been warned about this.
>
> (C2654 f 61 Birmingham)

> It is rather nerve-wracking to mix with people who are different, but I have generally found that once the initial greetings are over, the experience can be quite pleasant.
>
> Any time I have had to meet people from another country, the one thing that is so apparent is that we are all similar human beings under the skin, with a lot more in common that we realise.
>
> (C2053 f 50 Attleborough)

> his appearance was rather forbidding. He was a punk with a blonde mohican hairstyle. . . . The other groups in the class all gave us a wide berth and wouldn't sit anywhere near us. However, as I got to know J I began to realise what a gentle, kind and warm hearted young man he was, more concerned for others, and animals, than himself. I really enjoyed our lessons and it made me realise how important it was not to judge by appearances.
>
> (P1282 f 65 Lichfield, Staffs)

This is a long way from Young's (2002: 437) 'unoppressive city . . . defined as unassimilated otherness', which offers the potential for knowing others as different and gaining understanding of groups and cultures that are not one's own, where: 'In such public spaces the diversity of the city's residents come together and dwell side by side, sometimes appreciating one another, entertaining one another, or just chatting, always to go off as strangers.'

Richard Sennett (1996: 42) observed a similar trend emerging in American cities over the last quarter of the twentieth century. Neighbourhoods within cities, he suggested, are becoming increasingly homogeneous ethnically as people elect to live close to people like themselves in a drive towards a 'community of similarity'. As a result people lose the art of relating to and interacting with people who do not share the same language or understandings, so that they regard meeting and negotiating with others who are different with apprehension. This becomes a vicious circle which is hard to break. Bauman (2003a, 2003b) attributes this drive towards similarity and sameness to the fact that people in cities are overwhelmed and discomforted by the strangeness and unknowability of others who are different from themselves – echoing Simmel's notion of overstimulation leading to psychic retreat. Recalling Mumford's (1938) early analyses of smaller communities, Bauman suggests that alien others in villages and rural areas are enfolded into the community through being known and understood, through being domesticated and incorporated. In contrast, strangers in cities are too numerous to be familiarised – 'they are the unknown variable', whose intentions cannot be predicted and whose 'presence inside the fields of action is discomforting' (2003a: 27). Under these circumstances people have to make daily choices as to how to act, whether by design or default. Bauman suggests two opposing tendencies – 'mixophobia' and 'mixophilia' – both of which are present in the 'liquid modern city'. Mixophobia is a widespread response to the overwhelming variety of life styles and differences that people encounter each day in the city streets and to those: 'accumulated anxieties [which] tend to unload against the selected category of "aliens", picked up to epitomise "strangeness as such". In chasing them away from one's homes and shops, the frightening ghost of uncertainty is, for a time, exorcised' (2003a: 26). I came across this account after having already written about Princess Street market, which so sharply illustrates the sentiments expressed here. Like Sennett, and drawing on Caldeira's (1996: 303–328) research on São Paulo as an illustration, he suggests that mixophobia drives individuals into self-segregation in walled and fortified enclaves. Yet cities, in Bauman's view, as is strongly argued here, offer the potential for meeting strangers more intimately, for defusing the tension and apprehension that unknown others produce and for negotiating rules of life in common through co-operation and talk, or simply through eye contact – this is the potential for mixophilia.

RISK

Risk represents a related barrier to rubbing along in the public spaces of the city. As Furedi (1997: 9) argues, an evaluation of everything in terms of risk is the defining characteristic of contemporary society, where what is striking is not the level of insecurity, but rather the profoundly conservative way in which this condition is experienced. In his view risk has come to organise decisively the ways in which we conduct ourselves in everyday life:

> The disposition to perceive one's existence as being at risk has had a discernible effect on the conduct of life. It has served to modify action and interaction between people. The disposition to panic, the remarkable dread of strangers and the feebleness of relations of trust have all had important implications for everyday life. . . . Through the prism of the culture of abuse, people have been rediscovered as sad and damaged individuals in need of professional guidance. From this emerges the diminished subject, ineffective individuals and collectivities with low expectations. Increasingly we feel more comfortable with seeing people as victims of their circumstances rather than as authors of their lives. The outcome of these developments is a world view which equates the good life with self-limitation, and risk aversion.
>
> (ibid: 147)

Bauman sees some of this lack of trust in others as generated by the television programmes that are increasingly popular, particularly amongst the younger generation. In reality TV like *Big Brother* or *The Weakest Link*, the message is that we should regard each other with suspicion, that no one should be trusted and that humans are in competition with another and ultimately disposable (2003b: 87–88).

In this book we saw clearly how risk of litigation and blame was a salient motivation for the Corporation of London's move to curtail winter swimming in the ponds. For the older people in Chapter 6, the danger of public places, resulting either from the threat of violence from others, notably younger others, or from their being unsafe places for frail bodies, curtailed their sense of easy mobility. As

I argued there, the burden of risk reduction is increasingly forming part of the moral responsibility of urban citizens themselves (Rose 2000: 103) so that rather than space being made safe for older people, they are expected to minimise risk for themselves and withdraw into the private arena. Similarly parents' fears for their children limited their freedom to move in public spaces alone and outside adult supervision. Fear of change and the unknown, of what the future may bring, mobilising nostalgic reminiscence, as was illustrated in the London street market, also inhibited new ways of cohabiting with different others.

The challenge here is not to argue that cities and their public spaces are not also sites of risk, they can be. It is, however, to contest normative versions of risk which obscure and delimit social relations between people who are different from one another and which frighten categories of people off the street. Media representations of the all-pervasive paedophile waiting outside the school gates, or stories of gangs of youths, though occasionally 'true', are rarely challenged with accounts of what more realistic levels of care of the self and dependent others could be. As a result those who are frightened vacate public space, making it emptier and thus less safe as a consequence. Similarly various arenas of the state, as we saw here, deploy risk discourse to legitimate interventions in public space which are motivated from another source, or to abdicate responsibility for making a safer environment for urban citizens.

AFFECT EFFECT

Finally, public space itself has also come under attack from several directions – thematisation, enclosure into malls and other controlled spaces, and privatisation, or from urban planning and design interventions to erase its uniqueness, as we also saw in this book. Certainly the withdrawal of public space in its more traditional and known form has to be resisted. But it is also important to recognise that public spaces take many different forms, and the more visible and documented spaces of the mall or town square, the park or the piazza, though important, are only one part of the story. Symbolic or meaningful places may be hard to recognise. Marginal or liminal sites can often be more innovative and significant to people's lives – particularly to those more marginalised subjects in the city – as well as revealing of social norms. As Holston (1998: 54) puts it: 'Countersites are more than just indicators of the norm. They are themselves possible alternatives to it'. Moreover, the workings of normative power in the public spaces of the city, which are routine, expected, everyday, are often invisible or hard to detect, and thus hard to challenge or address. These spaces on the edge, symbolic space and spaces of the imaginary are more tricky to plan for, as are the more marginal and less powerful subjects of the city.

TOWARDS ENCHANTED URBAN ENCOUNTERS

My desire is for a city where public spaces are enchanted and inclusive, recognising also that these same spaces may also sometimes be spaces of exclusion and disenchantment shifting over different registers of temporality. Such an objective entails forms of planning that do not fix activity or use immutably, since all spaces contain the possibilities for openness and closure, and this will always be in a state of flux. Here Healey's (1997) argument for collaborative planning makes sense in that it recognises people's diverse interests and expectations, and reveals the implicit power relations, not just in the material distribution of resources, but in the fine-grained assumptions and practices of planning. Yet planning can only go so far in creating a city where people can encounter one another across their differences without closure, threat or violence. Agonistic relations are always entailed in relations and politics of difference, as Mouffe (1993: x), amongst others, is emphatic in demonstrating. Differences, in all their embodied and psychic complexities, will always enter and be performed in the public spaces of the city. As one difference is acknowledged and settled another will emerge in its place. Engagement across differences, a mutual respect for those who are different from oneself, and space for them to be so, is a precondition for space to be public. As Connolly argues:

> only a few liberals . . . pretend to leave their fundaments at home when they enter public space, and even then a lot of others see through the pretence. Everybody else openly brings large chunks of their bodies and souls with them into the public realm. The need today, then, is to forge a generous ethos of political engagement between multiple orientations, in which many participants come to terms affirmatively with the contestability of the moral faiths and identities they prize the most and in which many then embody that appreciation in the way they articulate, debate and decide fundamental issues of public life.
>
> (1998: 94)

Such arguments unsettle normative policy prescriptions which argue for community cohesion (Home Office 2001) based on a common vision and sense of belonging. Central to the Home Office intentions is the notion of building strong and positive relations between people of different backgrounds, and appreciating and valuing people from a diversity of backgrounds – who could argue with that? But the objective of bringing people together under a vision in common remains problematic, and common belonging in the near future may be an impossible dream – at least for as long as there are imbalances of power and a profound sense of injustice amongst many minorities in Britain. Equality between people of different racial backgrounds can only be achieved when racism, and its origins in Britain's imperial and colonial past, is eradicated (Gilroy 2004). In practical terms a more productive approach would be to build initiatives across the many spaces of multiple publics for overlap and cross-fertilisation in the context of the culturally heterogeneous society in which we now live (Amin 2002).

Difference is not just a way of describing the plurality of the social world we now inhabit, it is integral to the (post)modern self which is never complete and always fragmentary. Thus engagements in public arenas are also always temporary, contingent, partial, invented and reinvented, and open ended. For different selves, and different cultural practices to be articulated in the public realm is counter to normative liberal humanism, since people who do not conform to cultural norms, as in the case of the eruv, are told to leave their 'bags of faith at the door when they enter the public realm' (Connolly 1995: 124). Following Connolly I want to argue for a public realm which is more open-ended, contingent and elastic and which recognises that groups are never fixed but also always in some state of formation or dispersal. This implies that the state should actively cultivate in its citizens a sensibility which accepts cultural and social difference as the norm (ibid: 143). There are widespread implications here for intervention into the various arenas of the state, from education, social policy and planning to media, the arts and architecture.

Nancy Fraser (Fraser and Honneth 2003) similarly advocates a politics of recognition of others in their differences, albeit with a far less fluid sense of the self or group differences, while situating this politics of recognition centrally within a politics of redistribution as the other key tenet of social justice. This leads us to the broader socio-political context of this late or liquid modern period where global processes, a space of flows, are outside the control of any locale and generate a space of places that are localised, fragmented and often powerless (Castells 1996, 1997; Appadurai 1996; Short 2004). But at the same time, as part of the same process, the local becomes the site of struggles around identity and meaning, and even global players, those flying across the world on jet planes, or inhabiting cyberspace, are also located, looking for urban sites in which to live, play and intermingle with others. Meanwhile transnational flows of migration and cultural practices become enmeshed in the everyday lives and routines of local residents, conferring new socially constructed relations of power and meaning (Smith 2001: 1–3). The global has by no means consigned the local to history, indeed in some respects it has provided the very space for the local to be reasserted. Nevertheless, global processes have contributed to ever widening gaps between rich and poor countries and between the rich and the poor in most of the more developed countries across the world. Many citizens, particularly those who are migrants, those who are old, children, single parents, disabled and many other more marginal urban subjects, are disempowered and excluded. For citizens to assert and negotiate differences in public, their basic needs, including adequate shelter, income and employment (where relevant), must first be met.

In arguing for respect for others in their differences, as I do, I am also mindful of Sennett's (2004) recent book of that name – *Respect* – not Britain's current Prime Minister Tony Blair's invocation of this theme in his politics of blame, shame and punishment. Sennett's crucial point is that people need to feel respected themselves if they are to respect others, arguing that the collapse of traditional social institutions, limited job security for the many unskilled, and the economic drive towards flexibility has left many people feeling insecure. The restoration of respect in society requires the powerful and rich to show respect to those less powerful and skilled than themselves. In encounters across difference in the public spaces of the city, this same respect by the powerful and wealthy towards the powerless and marginal is a prerequisite in countering the movement to closure and withdrawal. Central to the ethics of respect is the need to tackle exclusion and poverty and central to a politics of recognition is redistribution (Fraser and Honneth 2003).

My argument centres on the need to address power relations on the one hand and, on the other, liberal humanistic values which construct and reinforce dominant publics, cultural imaginaries and practices while marginalising or subjugating others. The Citizenship Foundation espouses the following laudable aims:

> We see a multicultural society as one made up of a diverse range of cultures and identities, and one that emphasises the need for a continuous process of mutual engagement and learning about each other with respect, understanding and tolerance. . . . Such societies, under a framework of common civic values and common legal and political institutions, not only understand and tolerate diversities of identity but would also respect and take pride in them. . . . We do not imply that identities are fixed; in fact identities are often more fluid than many people suppose.
>
> (2003: 10)

Yet there remains the taxing question as to who is the 'we' in any such formulation, and who is doing the tolerating, which needs to be unpacked and subverted so that it embodies within its own discourses the very differences, agonistic, fluid, open-ended and contingent as these will be, that such a statement seeks to embrace. Public spaces are essential to such a politics and it is the argument of this book that the importance of public space for encountering different others, or even for encountering the other selves within (our selves), will mean recognising that those sites and spaces on the margins and those imaginary and symbolic spaces, or barely visible spaces, are quite as important as the monumental, well-designed and formal public spaces of the city – so popular with urban designers, planners and politicians. It is my hope that this book will contribute, in small part, to new enchantments of urban encounters in public space.

Appendix
A summary of the primary research methods

CHAPTER 2

Research on the two sites was conducted through a number of different methods. In the British case, Jewish and non-Jewish residents in Barnet, opposition residents' groups, four rabbis from different synagogues, local authority officials – including members of the Planning Department, the Road Authority and the superintendent of Hampstead Heath – were interviewed. Reports from the London Borough of Barnet Town Planning and Research Committee and the Public Works Committee, local and national newspaper articles, a television documentary and relevant websites were viewed, read and analysed. In the US case of Tenafly, the research was based on the report of the proceedings of United States Court of Appeals for the Third Circuit (No. 01–3301), Tenafly Residents Association Inc. against the Borough of Tenafly, a report by the Becket Fund for Religious Liberty, newspaper articles and relevant websites.

CHAPTER 3

The research for Chapter 3 took place between 2002 and 2003. Interviews were conducted with the assistance of Karen Wells in a street market in North London. Thirty in-depth interviews were conducted with stall holders, shopkeepers and council officials, including the regeneration officer, and several librarians from the local library located in the street. Unstructured observation was carried out in the market over several days and notes taken. Photographs provided further insights into daily social practices. Council documents and local area statistics were analysed.

CHAPTER 4

The research for Chapter 4 was carried out by me at the women's pond, and Mostafa Gamal at the men's pond, in 2003–2005. Interviews were conducted with pond users and members of the campaigns to save the ponds. Campaign and council meetings were attended and detailed notes taken. The social and cultural practices of pond users were observed and recorded over several days in each summer. Newspaper articles were collected for content and discourse analysis.

CHAPTER 5

The research for this chapter was carried out during 2003. In-depth interviews were carried out in six Turkish baths in London and three smaller cities. Mostafa Gamal visited the baths on men's days, and I carried out the research on the women's days. Detailed observation of social and cultural practices was recorded by the researchers. The attendants in charge of the baths were interviewed. A broader context for the study was provided by Malcolm Shifrin's website: www.victorianturkishbath.org. The material for the ponds section in this chapter was collected using the same methods as for Chapter 4.

CHAPTER 6

The research on U3A was carried out between 1999 and 2002. In-depth interviews with members of U3A were conducted in several London boroughs, and a number of U3A groups outside the metropolis, several from small towns in rural England (north and south). Material from the U3A website was also analysed. The in-depth interviews with allotment holders and older gay people (a total of ten in each group) were carried out by John Austin. A focus group was held with a group of older people in a London library. A focus group was conducted with six members of Better Government for Older People.

CHAPTER 7

Research was carried out in two London primary schools during 2003. Karen Wells conducted the South London research and I conducted the North London research. Access to the schools was obtained through the heads of the schools and consent was given by parents to interview the children. A class of ten-year-olds in each school was selected. Several methods were used: the children were asked to draw pictures of the public places they liked, disliked and visited most often in their locality. The children were given disposable cameras to take home and asked to take pictures of public places that were significant to them. The children returned to the class with their cameras and the photographs were developed. Small focus groups were held with the children in both schools where the pictures and photographs were used as prompts to the discussion. The photographs and drawings were also analysed by me and categorised according to content and themes.

CHAPTER 8

The Mass Observation data were derived by placing questions on private and public space and difference, for discussion in the Autumn Directive (70) 2003. Respondents to the archive are asked to write their thoughts and ideas on the subjects outlined in the directive. These are anonymous; 180 respondents replied. The material was thematically coded.

Bibliography

Adams, J. (1995) *Risk*, London: UCL Press

Ahmed, S. (2000) *Strange Encounters: Embodied Others in Post-coloniality*, London: Routledge

Aldrich, R. (2004) Homosexuality and the City: An Historical Overview, *Urban Studies*, vol. 41, no. 9, pp. 1719–1737

Allen, J. (2003) *Lost Geographies of Power*, Oxford: Blackwell

Allen, J., Massey, D. and Pryke, M. (eds) (1999) *Unsettling Cities: Movement and Settlement*, London and New York: Routledge and Open University Press

Altman, D. (2001) *Global Sex*, Chicago, IL: University of Chicago Press

Amateur Swimming Association (ASA) (2003) *Report Commissioned by the Corporation of London on Winter Swimming*, London

American, S. (1898) The Movement for Small Playgrounds, *American Journal of Sociology*, vol. 2, no. 4, pp. 159–170

Amin, A. (2002) *Ethnicity and the Multicultural City: Living with Diversity*, Report for the Department of Transport, Local Government and the Regions and the ESRC Cities Initiative, London

Amin, A. and Thrift, N. (2002) *Cities: Re-imagining the Urban*, Cambridge: Polity

Anderson, A. (2001) *The Powers of Distance: Cosmopolitanism and the Cultivation of Detachment*, Princeton, NJ: Princeton University Press

Anderson, B. (1996) Introduction, in B. Balakrishnan (ed.) *Mapping the Nation*, London: Verso

Anderson, E. (1990) *Streetwise: Race, Class and Change in an Urban Community*, Chicago, IL: Chicago University Press

Appadurai, A. (1996) *Modernity at Large*, Minneapolis, MN and London: University of Minnesota Press

Arendt, H. (1958) *The Human Condition*, Chicago, IL: University of Chicago Press

Ariès, P. (1962) *Centuries of Childhood: A Social History of Family Life*, New York: Random House

Back, L., Crabbe, T. and Solomos, J. (2001) *The Changing Face of Football: Racism, Identity and Multiculture in the English Game*, Oxford: Berg

Balakrishnan, B. (ed.) (1996) *Mapping the Nation*, London: Verso

Barrett, M. (1980) *Women's Oppression Today*, London: Verso

Bartlett, S. (2002) Building Better Cities with Children and Youth, *Environment and Urbanisation*, vol. 14, no. 2, pp. 3–10

Bauman, Z. (2003a) *City of Fears, City of Hopes*, London: Goldsmiths College, Centre for Urban and Community Research

Bauman, Z. (2003b) *Liquid Love: On the Frailty of Human Bonds*, Cambridge: Polity
Beauregard, R. (1993) *Voices of Decline*, Oxford: Blackwell
Beazley, H. (2000) Street boys in Yogyakarta: Social and Spatial Exclusion in the Public Spaces of the City, pp. 472–488 in G. Bridge and S. Watson (eds) *A Companion to the City*, Oxford: Blackwell
Beck, U. (1992) *Risk Society*, London: Sage
Becket Fund (2002) Tenafly Eruv Association, Inc. v. The Borough of Tenafly (www.becketfund.org/litigate/Tenafly.html)
Bell, D. and Binnie, J. (2004) Authenticating Queer Space: Citizenship, Urbanism and Governance, *Urban Studies*, vol. 41, no. 9, pp. 1807–1820
Benhabib, S. (2002) *The Claims of Culture: Equality and Diversity in a Global Era*, Princeton, NJ: Princeton University Press
Benjamin, W. (1999) The Arcades Project, transl. H. Eiland and K. McLaughlin pp. 393–400 in G. Bridge and S. Watson (2002) *The Blackwell City Reader*, Oxford: Blackwell
Bennett, J. (2001) *The Enchantment of Modern Life: Attachments, Crossings and Ethics*, Princeton, NJ: Princeton University Press
Bernstein, R. J. (1996) *Hannah Arendt and the Jewish Question*, Cambridge: Polity
Betz, H. G. (1994) *Radical Right-Wing Populism in Western Europe*, London: Macmillan
Blackford, H. (2004) Ring-Around-the-Children, a Pocketful of Women, *Childhood*, vol. 11, no. 2, pp. 227–249
Blank, J. (ed.) (2000) *Still Doing It. Women and Men over 60 Write about Their Sexuality*, San Francisco, CA: Down There Press
Borden, I. (2001) *Skateboarding: Space and the City*, Oxford: Berg
Boyd, A. (2003) Turkish Baths, *Whistler Centre Journal*, 17 December
Brauner, L. S. (2002) Appellate Court Allows Tenafly Eruv to Remain Up, *Chicago Jewish Community on Line* (Jewish United Fund News and Public Affairs)
Bridge, G. and Watson, S. (eds) (2000) *A Companion to the City*, Oxford: Blackwell
Brown, W. (2005) Tolerance As/In Civilizational Discourse, Lecture, Open University, Launch of Centre for Citizenship, Identities and Governance
Brue, A. (2003) *Cathedrals of the Flesh: My Search for the Perfect Bath*, London: Bloomsbury
Buck-Morss, S. (1989) *Dialectics of Seeing*, Boston, MA: MIT Press
Bullock, R. and Gould, G. (1988) *The Allotment Book*, London: MacDonald.
Butcher, G. W. (1918) *Allotments for All. The Story of a Great Movement*, London: Allen and Unwin
Butler, J. (1993) *Bodies that Matter*, London: Routledge
Butler, J. (1997) *The Psychic Life of Power*, Stanford, CA: Stanford University Press
Caldeira, T. (1996) Fortified Enclaves: The New Urban Segregation, *Public Culture*, pp. 303–328
Callard, F. J. (1998) The Body in Theory, *Environment and Planning, D. Society and Space*, 16, pp. 387–400
Cane, M. and Griswold, A. (2002) *The Hungry Swimmer*, London: Kenwood Ladies Bathing Association
Carr, S., Francis, M., Rivlin, L. G. and Stone, A. M. (1992) *Public Space*, Cambridge: Cambridge University Press
Castells, M. (1996) *The Rise of the Network Society*, Oxford: Blackwell
Castells, M. (1997) *The Power of Identity*, Oxford: Blackwell

Chase, M. and Shaw, C. (eds) (1989) *The Imagined Past: History and Nostalgia*, Manchester: Manchester University Press

Chauncey, G. (1994) *Gay New York: Gender and Urban Culture and the Making of the Gay Male World 1890–1940*, New York: Basic Books

Cheah, P. and Robbins, B. (eds) (1998) *Cosmopolitics*, Minneapolis, MN: University of Minnesota Press

Citizenship Foundation (2003) *The New and the Old*, Report of the Life in the United Kingdom Advisory Group

Cloke, P. and Jones, O. (2005) 'Unclaimed Territory': Childhood and Disordered Space(s), *Social and Cultural Geography,* vol. 6, no. 3, June, pp. 313–333

Colley, L. (2003) *Britons: Forging the Nation, 1707–1837*, London: Pimlico

Collins, A. (2004) Sexual Dissidence, Enterprise and Assimilation: Bedfellows in Urban Regeneration, *Urban Studies*, vol. 41, no. 9, pp. 1789–1806

Connolly, W. E. (1995) *The Ethos of Pluralisation*, Minneapolis, MN: University of Minnesota Press

Connolly, W. E. (1998) Rethinking the Ethos of Pluralisation, *Philosophy and Social Criticism*, vol. 24, no. 1, pp. 93–102

Connolly, W. E. (2002) *Identity/Difference Democratic Negotiations of Political Paradox*, Minneapolis, MN: University of Minnesota Press

Cooper, D. (1996) Talmudic Territory? Space, Law and Modernist Discourse, *Journal of Law and Society*, vol. 23, no. 4, December, pp. 529–548

Cooper, D. (1998) *Governing out of Order*, Cambridge: Polity

Copjec, J. and Sorkin, M. (eds) (1999) *Giving Ground: The Politics of Propinquity*, London: Verso

Crang, M. and Thrift, N. (eds) (2000) *Thinking Space*, London: Routledge

Crawford, M. (1992) The World in a Shopping Mall, pp. 3–30 in Michael Sorkin (ed.) *Variations on a Theme Park: The New American City and the End of Public Space*, New York: Hill and Wang

Crouch, D. and Ward, C. (1988) *The Allotment. Its Landscape and Culture*, London: Faber and Faber

Curry, H. (2002) The Heat Is On, *Guardian*, 7 September 2002

Davis, F. (1979) *Yearning for Yesterday: A Sociology of Nostalgia*, New York and London: The Free Press

Davis, M. (1992) *City of Quartz*, New York: Vintage

Davis, M. (1998) *Ecology of Fear: Los Angeles and the Imagination of Disaster*, New York: Metropolitan Books

Davis, M. (2002) *Dead Cities and Other Tales*, New York: New Press

Deakin, R. (2000) *Waterlog: A Swimmer's Journey through Britain*, London: Vintage

Dean, M. (1999) *Governmentality: Power and Rule in Modern Society*, London: Sage

de Beauvoir, S. (1972) *Old Age*, transl. P. O'Brian, London: André Deutsch and Weidenfeld and Nicolson

de Certeau, M. (1984) *The Practice of Everyday Life*, Berkeley, CA: University of California Press

Delany, S. (1999) *Times Square Red, Times Square Blue*, New York: New York Press

Deutsche, R. (1996) *Evictions: Art and Spatial Politics*, Boston, MA: MIT Press

Deutsche, R. (1999) Reasonable Urbanism, pp.175–206 in J. Copjec and M. Sorkin (eds) *Giving Ground: The Politics of Propinquity*, London: Verso

Douglas, M. (1991) Witchcraft and Leprosy: Two Strategies of Exclusion, *Man*, vol. 26, pp. 723–736

Douglas, M. and Wildavsky, A. (1983) *Risk and Culture: An Essay on the Selection of Technological and Environmental Dangers*, Berkeley, CA: University of California Press

Elkin, J. and Kinnell, M. (2000) *A Place for Children*, London: Library Association Publishing

Elsley, S. (2004) Children's Experience of Public Space, *Children and Society*, vol. 18, no. 2, pp. 155–164

Emecheta, B. (1987) *Second-Class Citizen,* London: Fontana Paperbacks (first published 1974 by Allison and Busby, London)

Ennew, J. P. (1994) Time for Children or Time for Adults?, pp. 125–143 in J. Qvortrup, M. Brady, G. Sgritta and H. Wintersberger (eds) *Childhood Matters: Social Theory, Practice and Politics*, Aldershot: Avebury

Etoe, C. (2000) *Vandals Bring Holiday Misery*, *CNJ*, 27 July, p. 4

Eyres, H. (2004) Quality Sacrified to Meanness, *Financial Times*, p. 20

Felski, R. (1995) *The Gender of Modernity*, Boston, MA: Harvard University Press

Foucault, M. (1988) *The Care of the Self*, New York: Vintage

Fraser, N. (1998) *Redistribution or Recognition? A Political-Philosophical Exchange* (with A. Honneth), London: Verso

Fraser, N. and Honneth, A. (2003) *Redistribution or Recognition*, London: Verso

Friends of the Earth (1974) *The Allotments Campaign Manual*, London: FoE

Furedi, F. (1997) *Culture of Fear: Risk-taking and the Morality of Low Expectation*, London: Cassell

Futerman, V. (ed.) (1989) *Into the 21st Century*, Cambridge: Symposium Committee of U3A

Gale, R. and Ryan, J. R. (2002) Religion, Planning and the City: The Spatial Politics of Ethnic Minority Expression in British Cities and Towns, *Ethnicities*, vol. 2, no. 3, pp. 387–409

Gallagher, C. (2004) Children as Advocates for Change in the City, *Childhood*, vol. 11, no. 2, pp. 251–262

Geertz, C. (1973) *The Interpretation of Cultures*, New York: Basic Books

Gelder, K. and Jacobs, J. (1998) *Uncanny Australia*: *Sacredness and Identity in a Post-colonial Nation*, Melbourne: Melbourne University Press

Giddens, A. (1990) *Consequences of Modernity*, Cambridge: Polity

Giddens, A. (1994) *Beyond Left and Right: The Future of Radical Politics*, Cambridge: Polity

Giddens, A. (1998) Risk Society: The Context of British Politics, pp. 23–34 in J. Franklin (ed.) *The Politics of Risk Society*, Cambridge: Polity Press

Giddens, A. (1999) *Runaway World: How Globalisation Is Reshaping our Lives*, London: Profile Books

Gilbraith, A. (1998) Corporation of London: Eruv is Threat to Heath, *Ham and High* 26 June, p. 5

Gilroy, P. (2004) *After Empire: Melancholic or Convivial Culture*, London: Routledge

Goffman, E. (1963) *Behavior in Public Places*, New York: Free Press

Goffman, E. (1971) *Relations in Public: Microstudies of the Public Order*, New York: Basic Books

Goldstein, J. H. (ed.) (1994) *Toys, Play and Child Development*, Cambridge: Cambridge University Press

Gregson, N. and Rose, G. (2000) Taking Butler Elsewhere: Performativities, Spatialities and Subjectivities, *Society and Space*, vol. 18, pp. 433–452

Griswold, A. (1998) *Kenwood Ladies Bathing Pond*, London: KLBA

Groombridge, B. (1995) Emergent Challenges for Universities of the Third Age, pp. 27–37 in E. Heikkinen, J. Kuusinen and I. Ruoppila (eds) *Preparing for Aging*, New York: Plenum Press

Grosz, E. (1992) Bodies – Cities, pp. 241–254 in Beatrice Colomina (ed.) *Sexuality and Space*, Princeton, NJ: Princeton University Press

Grosz, E. (1995) *Volatile Bodies: Towards a Corporeal Terminism*, Sydney: Allen and Unwin

Gutmann, A. (ed.) (1998) *Freedom of Association*, Princeton, NJ: Princeton University Press

Habermas, J. (1989) *The Structural Transformation of the Public Sphere: An Inquiry into a Category of Bourgeois Society*, Cambridge, MA: MIT Press

Hage, G. (1998) *White Nation*, Sydney: Pluto Press

Hall, S. (2000) The Multicultural Question, pp. 209–241 in B. Hesse (ed.) *Un/settled Multiculturalisms*, London: Zed Books

Hamnett, C. (2000) Gentrification, Post-industrialism and Restructuring, pp. 331–341 in G. Bridge and S. Watson (eds) *A Companion to the City*, Oxford: Blackwell

Hart, R. (1979) *Children's Experience of Place*, New York: Irvington

Hart, R. (2002) Containing Children: Some Lessons on Planning for Play from NYC, *Environment and Urbanisation,* vol. 14, no. 2, pp. 135–148

Hayden, D. (1996) What Would a Non-sexist City Look Like? Speculations on Housing, Urban Design and Human Work, pp. 448–464 in R. LeGates and F. Stout (eds) *The City Reader*, London: Routledge

Healey, P. (1997) *Collaborative Planning: Shaping Places in Fragmented Societies*, London: Macmillan

Healey, P. (2000) Planning in Relational Space and Time, pp. 517–530 in G. Bridge and S. Watson (eds) *A Companion to the City*, Oxford: Blackwell

Henaff, M. and Strong, T. B (eds) (2001) *Space and Democracy*, Minneapolis, MN: University of Minnesota Press.

Holston, J. (1998) Spaces of Insurgent Citizenship, in L. Sandercock (ed.) *Making the Invisible Visible,* Los Angeles, CA: University of California Press

Home Office (2001) *Community Cohesion* (The Cantle Report), London: HMSO

Hood, C., Rothstein, H. and Baldwin, R. (2004) *The Government of Risk*, Oxford: Oxford University Press

Horwood, C. (2000) 'Girls who Arouse Dangerous Passions': Women and Bathing, 1900–39, *Women's History Review*, vol. 9, no. 4, pp. 653–673

Howe, J. (2000) Planning for Sustainable Communities: The Case for Urban Food. Doc. 2 Summary ESRC Grant no. (R000222844). Dept of Planning and Landscape. University of Manchester

Hubbard, P. (2002) Sexing the Self: Geographies of Engagement and Encounter, *Social and Cultural Geography*, vol. 3, no. 4 pp. 365–381

Huber, J. and Skidmore, P. (2003) *Why the Baby Boomers Won't Be Pensioned Off*, London: Demos and Age Concern

Humphries, S., Mack, J. and Perks, R. (1988) *A Century of Childhood*, London: Sidgwick and Jackson

Iveson, K. (2003) Justifying Exclusion: The Politics of Public Space and the Dispute over Access to McIvers Ladies Baths, Sydney, *Gender, Place and Culture*, vol. 19, no. 3, pp. 215–228

Jackson, J. B (1984) *Discovering the Vernacular Landscape*, Newhaven, Conn.: Yale University Press

Jacobs, J. (1966) *Death and Life of American Cities*, London: Jonathan Cape

Jacobus, M. (1995) *First Things: The Maternal Imaginary in Literature, Art and Psychoanalysis*, London: Routledge

Jans, M. (2004) Children as Citizens: Towards a Contemporary Notion of Child Participation, *Childhood*, vol. 11, no. 1, pp. 27–44

Jenkins, R. (2000) Disenchantment, Enchantment and Re-enchantment: Max Weber at the Millennium, Max Weber Studies, vol. 1, no. 1, pp. 11–32

Karsten, L. (2003) Children's Use of Public Space, *Childhood*, vol. 10, no. 4, pp. 457–473

Katz, C. (1995) Power, space and terror: social reproduction and the public environment, Paper presented at the Landscape, Architecture, Social Ideology and the Politics of Place conference, Cambridge, MA, Harvard University

Katz, S. (1996) *Disciplining Old Age: The Formation of Gerontological Knowledge*, Charlottesville: University Press of Virginia

Keith, M. and Pile, S. (eds) (1993) *Place and the Politics of Identity*, London: Routledge

Kiss, E. (1999) Democracy and the Politics of Recognition, pp. 193–209 in J. Shapiro and C. Hacker-Cordon (eds) *Democracy's Edges*, Cambridge: Cambridge University Press

KLPA (2002) *The Hungry Swimmer*, London: KLPA

Kristeva, J. (1991) *Strangers to Ourselves*, transl. Leon Roudiez, New York: Columbia University Press

Laclau, E. (2000) Identity and Hegemony. The Role of Universality in the Constitution of Political Logics, pp. 44–89 in J. Butler, E. Laclau and S. Zizek, *Contingency, Hegemony, Universality: Contemporary Dialogues on the Left*, London: Verso

Laclau, E. and Mouffe, C. (1985) *Hegemony and Socialist Strategy*, London: Verso

Laslett, P. (1995) The Third Age and the Disappearance of Old Age, in E. Heikkinen, J. Kuusinen and I. Ruoppila (eds) (1995) *Preparation for Aging*, New York: Plenum Press.

Laslett, P. (1996) *A Fresh Map of Life*, London: Weidenfeld and Nicolson

Lefebvre, H. (1991) *The Production of Space*, transl. D. Nicholson-Smith, Oxford: Blackwell

Lefort, C. (1988) *Democracy and Political Theory*, Minneapolis, MN: University of Minnesota Press

Levinas, E. (1994) The Rights of Man and the Rights of the Other, pp. 116–125 in *Outside the Subject*, transl. M. B. Smith, Stanford, CA: Stanford University Press

Levinas, E. (1999) *Alterity and Transcendence*, transl. M. B. Smith, London: Athlone Press

London Borough of Barnet (1993) *Town Planning and Research Committee Report*, Item no. 4, 27 October

Longhurst, R. (2001) *Bodies: Exploring Fluid Boundaries*, London: Routledge

Lowenthal, D. (1989) Nostalgia Tells It Like It Wasn't, in C. Shaw and M. Chase (eds) *The Imagined Past: History and Nostalgia*, Manchester: Manchester University Press

Lynch, K. (ed.) (1977) *Growing up in Cities. Studies of the Spatial Environment of Adolescence in Cracow, Melbourne, Mexico City, Salta, Toluca, Warszwa*, Cambridge, MA and London: MIT Press

Madanipour, A. (2003) *Public and Private Spaces of the City*, London: Routledge

Marcuse, P. (1995) Not Chaos but Walls: Postmodernism and the Partitioned City, pp. 243–253 in S. Watson and K. Gibson (eds) *Postmodern Cities and Space*, Oxford: Blackwell

Marcuse, P. (2000) *Cities in Quarters*, pp. 270–281 in G. Bridge and S. Watson (eds) *A Companion to the City*, Oxford: Blackwell

Markusen, A. (2004) The Work of Forgetting and Remembering Places, *Urban Studies*, vol. 41, no. 12, pp. 2302–2313

Marx, K. (2003) On the Jewish Question, in L. Alcoff and E. Mendieta (eds) *Identities Race, Class, Gender and Nationality*, Oxford: Blackwell

Massey, D. (1995) Thinking Radical Democracy Spatially, *Environment and Planning, D Society and Space*, vol. 13, pp. 283–288

Massey, D. (1999) Entanglements of Power: Reflections, pp. 279–286 in J. Sharp, P. Routledge, C. Philo and R. Paddison (eds) *Entanglements of Power*, London: Routledge

Matthews, H. and Limb, M. (1999) Defining an Agenda for the Geography of Children: Review and Prospect, *Progress in Human Geography*, vol. 23, no. 1, pp. 61–90

Matthews, H., Limb, M. and Percy-Smith, B. (1998) Changing Worlds: The Microgeographies of Young Teenagers, *Tijdschrift voor Economische en Sociale Geografie*, 89, no. 2, pp. 193–202

Matthews, H., Limb, M. and Taylor, M. (1998) *Young People's Participation and Representation in Society*, University College, Northampton, Centre for Children and Youth

May, J. (1996) Globalisation and the Politics of Place: Place and Identity in an Inner London Neighbourhood, *Transactions*, vol. 21, pp. 194–215

May, J. and Thrift, N. (eds) (2001) *Timespace: Geographies of Temporality*, London: Routledge

Midwinter, E. (1992) *Citizenship: From Ageism to Participation*. The Carnegie Enquiry into the Third Age, Research paper no. 8, London: Carnegie Trust and Center for Policy on Ageing

Midwinter, E. (2004) *500 Beacons: The U3A Story*, London: Third Age Press

Mitchell, D. (2003) *The Right to the City*, New York and London: The Guilford Press

Modood, T. (2003) Muslims and European Multiculturalism, openDemocracy.net

Morris, S. (2002) Testing the Boundaries of Faith, *Guardian,* 10 August, p. 8

Mort, F. (2000) The Sexual Geography of the City, pp. 307–315 in G. Bridge and S. Watson (eds) *A Companion to the City*, Oxford: Blackwell

Mouffe, C. (1993) *The Return of the Political,* London: Verso

Mouffe, C. (2000) *The Democratic Paradox*, London: Verso

Mumford. L. (1938) *The Culture of Cities*, London: Martin Secker and Warburg

Mythen, G. (2004) *Ulrich Beck: A Critical Introduction to the Risk Society*, London: Pluto Press

Nancy, J. L. (1993) *The Birth to Presence*, Stanford, CA: Stanford University Press

Naylor, S. and Ryan, J. R. (2002) The Mosque in the Suburbs: Negotiating Religion and Ethnicity in South London, *Social and Cultural Geography*, vol. 3, no. 1, pp. 39–59

Newson, J. and Newson, E. (1976) *Four Years Old in an Urban Community*, Harmondsworth: Penguin

O'Brien, M., Jones, D., Sloan, D. and Rustin, M. (2000) Children's Independent Spatial Mobility in the Urban Public Realm', *Childhood: Special Issue: Spaces of Childhood*, vol. 7, no. 3, pp. 243–256

Parekh, B. (2000) *Rethinking Multiculturalism*, Basingstoke: Palgrave

Pateman, C. (1988) *The Sexual Contract*, Cambridge: Polity Press

Pile, S. and Keith, M. (eds) (1997) *Geographies of Resistance*, London: Routledge
Pile, S. and Thrift, N. (eds) (2000) *City A–Z: Urban Fragments*, London: Routledge
Pringle, R. and Watson, S. (1992) 'Women's Interests and the Post-structural State' in M. Barrett and A. Phillips (eds) *Destabilising Theory*, Oxford: Polity Press
Proust, M. (1983) *Remembrance of Things Past*, vol. 1: *Swann's Way*, transl. Scott Moncrieff and T. Milmartin, Harmondsworth: Penguin
Purdy, M. (2001) Our Towns: A Wire-Thin Line Sharply Divides a Suburb's Jews, *New York Times,* 25 March, Section 1, p. 35, col. 2
Putnam, R. (1993) *Making Democracy Work,* Princeton, NJ: Princeton University Press
Putnam, R. D. (2000) *Bowling Alone: The Collapse and Revival of American Community,* New York: Simon Schuster
Riccio, B. (1999) Senegalese Street-sellers, Racism and the Discourse on 'irregular trade' in Rimini, *Journal of Modern Italy*, vol. 4, no. 2, pp. 225–239
Riggio, E. (2002) Child Friendly Cities: Good Governance in the Best Interests of the Child, *Environment and Urbanisation*, vol. 14, no. 2, pp. 45–58
Robbins, B. (ed.) (1993) *The Phantom Public Sphere*, Minneapolis, MN: University of Minnesota Press
Robertson, R. (1999) *The Jewish Question in German Literature 1749–1939*, Oxford: Oxford University Press
Rofes, E. (2001) Imperial New York: Destruction and Disneyfication under Emperor Giuliani, *GLQ:* A *Journal of Lesbian and Gay Studies*, vol. 7, pp. 101–109
Rose, N. (2000) Governing Cities, Governing Citizens, pp. 95–109 in E. Isin (ed.) *Democracy, Citizenship and the Global City*, London: Routledge
Ryan, A. (1998) *The City and Free Association*, pp. 314–329 in A. Gutmann (ed.) *Freedom of Association*, Princeton, NJ: Princeton University Press
Samuel, R. (1989) *Patriotism: The Making and Unmaking of British National Identity*, vol. 11, London: Routledge
Sandercock, L. (1998) *Towards Cosmopolis*, Chichester: Wiley
Saunders, P. (1990) *A Nation of Home Owners*, London: Unwin Hyman
Seaford, H. (2001) Children and Childhood: Perceptions and Realities, *Political Quarterly*, vol. 72, no. 4, pp. 454–465
Seltzer, R. M. (1980) *Jewish People/Thought*, London: Macmillan
Sennett, R. (1970) *The Uses of Disorder*, Harmondsworth: Penguin
Sennett, R. (1974) *The Fall of Public Man*, New York: Norton
Sennett, R. (1990) *The Conscience of the Eye: The Design of Social Life of Cities*, London: Faber
Sennett, R. (1994) *Flesh and Stone: The Body and City in Western Society*, New York: Norton
Sennett, R. (1996) *The Uses of Disorder: Personal Identity and City Life*, London: Faber and Faber
Sennett, R. (2000) *Reflections on the Public Realm*, pp. 380–387 in G. Bridge and S. Watson (eds) *A Companion to the City*, Oxford: Blackwell
Sennett, R. (2004) *Respect: The Formation of Character in an Age of Inequality*, New York and London; Norton
Sharp, J., Routledge, P., Philo, C. and Paddison, R. (eds) (1999) *Entanglements of Power*, London: Routledge
Shaw, C. and Chase, M. (eds) (1989) *The Imagined Past: History and Nostalgia*, Manchester: Manchester University Press

Shields, C. (2002) *Unless*, London: Fourth Estate

Shifrin, M. R. (2004) Victorian Turkish Baths, www.victorianturkishbath.org, April

Short, J. R. (2004) *Global Metropolitan: Globalising Cities in a Capitalist World*, London: Routledge

Sibalis, M. (2004) Urban Space and Homosexuality: The Example of the Marais, Paris' Gay Ghetto, *Urban Studies*, vol. 41, no. 9, pp. 1739–1758

Sibley, D. (1995) *Geographies of Exclusion: Society and Difference in the West*, London: Routledge

Sieghart, M. (2004) Reclaim the Streets for Children, *The Times*, 3 March, *T2*, p. 3

Simmel, G. [1903] (2002) *Metropolis and Mental Life*, pp. 11–20 in G. Bridge and S. Watson (eds) *A Companion to the City*, Oxford: Blackwell

Smith, M. P. (2001) *Transnational Urbanism: Locating Globalisation*, Oxford: Blackwell

Smith, N. (1996) *The New Urban Frontier: Degentrification and the Revanchist City*, London: Routledge

Smithson, H. (2000) War of the Words, *Ham and High*, 4 August, p. 29

Sorkin, M. (1992) *Variations on a Theme Park: The New American City and the End of Public Space*, New York: Hill and Wong

Sources (2004) An Education Bulletin, www.u3a.org.uk

Staeheli, L. (1996) Publicity, Privacy and Women's Political Action, *Society and Space*, vol. 14, pp. 601–619

Stallybrass, P. and White, A. (1986) *The Politics and Poetics of Transgression*, London: Methuen

Steinwand, J. (1997) The Future of Nostalgia in Friedrich Schlegel's Gender Theory: Casting German Aesthetics Beyond Ancient Greece and Modern Europe, in J. Pickering and S. Kehde (eds) *Narratives of Nostalgia, Gender and Nationalism*, London: Macmillan

Sutton-Smith, B. (1994) Does Play Prepare for the Future?, pp. 130–147 in J. H. Goldstein (ed.) *Toys, Play and Child Development*, Cambridge: Cambridge University Press

Third Age Trust National Executive Committee (TATNEC) (2001) *U3A Membership Survey*, DfEE Contract Ref: IL 1992000/001 (sample 3,034 members)

Thompson, J. (1993) *Public Space* by S. Carr, book review, *Children's Environments*, vol. 10, no. 2

Tuan, Y. F. (1974) *Topophilia: A Study of Environmental Perception, Attitudes and Values*, Englewood Cliffs, NY: Prentice Hall

Tudor, A. (2003) (Macro) Sociology of Fear, *Classical Sociology*, vol. 51, no. 2, pp. 238–56

Valentine, G. (1996) Angels and Devils: Moral Landscapes of Childhood, *Environment and Planning D, Society and Space*, vol. 14, pp. 581–599

Valentine, G. (1997) 'Oh Yes I Can.' 'Oh No You Can't': Children and Parents' Understandings of Kids' Competence to Negotiate Public Space Safely, *Antipode*, vol. 29, no. 1, pp. 65–89

Valins, O. (2000) Institutionalised Religion: Sacred Texts and Jewish Spatial Practice, *Geoforum*, vol. 31, pp. 575–586

Villa, D. R (2001) Theatricality in the Public Realm of Hannah Arendt, pp. 144–171 in M. Henaff and T. B. Strong (eds) *Public Space and Democracy*, Minneapolis, MN: University of Minnesota Press

Vincent, P. and Warf, B. (2002) Eruvim: Talmudic Places in a Postmodern World, *Transactions of the Institute of British Geographers*, NS vol. 27, pp. 30–51

Waiko, J. (1992) Tugata: Culture, Identity, and Commitment, pp. 233–266 in L. Foerstal and A. Gilliam (eds) *Confronting the Margaret Mead Legacy: Scholarship, Empire and the South Pacific*, Philadelphia, Pa.: Temple University Press

Walker, B. (1997) *Sexuality and the Elderly. A Research Guide*, Westport, Conn., and London: Greenwood Press

Walzer, M. (ed.) (1995) *Toward a Global Civil Society*, Providence, RI: Berghahn Books

Ward, C. (1978) *The Child in the City*, Bedford: Square Press.

Watson, S. (1986) *Housing and Homelessness: A Feminist Perspective*, London: Routledge

Watson, S. (ed.) (1989) *Playing the State: Australian Feminist Interventions*, London: Verso

Watson, S. and Studdert, D. (2006) *Spaces of Diversity: Markets as Shared Social Spaces*, York: Joseph Rowntree Foundation

Wells, K. and Watson, S. (2005) A Politics of Resentment: Shopkeepers in a London Neighbourhood, *Ethnic and Racial Studies*, vol. 28, no. 2, pp. 261–277

Wildavsky, A. (1988) *Searching for Safety*, New Brunswick, NJ: Transaction

Williams, R. (1973) *The Country and the City*, London: Chatto and Windus

Williams, R. (1975) *The Country and the City*, Oxford: Oxford University Press

Wilmott, P. and Young, M. (1957) *Family and Kinship in East London*, London: Routledge

Wiltshire, R. and Crouch, D. (2001) Sustaining the Plot: Communities, Gardens and Land Use, *Town and Country Planning*, vol. 70, no. 9, p. 235

Wiltshire, R., Crouch, D. and Azuma, A. (2000) Contesting the Plot: Environmental Politics and the Urban Allotment Garden in Britain and Japan, pp. 203–217 in P. Stott and S. Sullivan (eds) *Political Ecology: Power Myth and Science*, London: Edward Arnold

Winnicott, D. (1971) *Playing and Reality*, London: Tavistock Publications

Wlakowitz, J. (1992) *City of Dreadful Delight: Narratives of Sexual Danger in Late Victorian London*, London: Virago

Worpole, K. (2000) *Here Comes the Sun: Architecture and Public Space in 20th Century European Culture*, London: Reakton Books

Young, I. (1990) *Justice and the Politics of Difference*, Princeton, NJ: Princeton University Press

Young, I. (2002) The Ideal of Community and the Politics of Difference, *Social Theory and Practice*, vol. 12, no. 1, pp. 430–439

Young, M. and Schuller, T. (1991) *Life After Work*, London: HarperCollins

Index